PREFACE

Applications of Probability and Statistics in Health Physics was the topic for the year 2000 Summer School sponsored by the Health Physics Society. It was held on the campus of Colorado State University in Ft. Collins, CO.

The approach for this school was not based on rigorous mathematics. Emphasis was placed on quantitative analysis using realistic situations rather than on formal derivation of theorems. Most of the topics were presented in a workshop atmosphere where students could use their portable computers to perform computations along with the faculty. This placed a restriction on the number of topics and the extent to which each could be discussed.

The Academic Dean made the selection of topics with the unfortunate realization that some very important subjects would have to be omitted. The idea was to include traditional methods as well as provide the opportunity to discuss contemporary approaches to relevant statistical problems.

This book served as a guide for students enrolled in the school but might also be useful for other persons interested in this subject. Each chapter was written by the faculty members who prepared and delivered the lectures. The first two chapters are meant to serve as an introduction. They could be considered as refresher courses on basic concepts and terminology for both classical and Bayesian analysis. The remaining chapters are each devoted to a specific topic in applied probability and statistics with examples relating directly to health physics.

The success of any book depends on the authors. For this there was no compromise. The authors are all distinguished scholars. But in addition, they are experienced teachers. It is an honor for me to be associated with them.

The topic of probability and statistics is often intimidating. Many persons find the concepts difficult to understand. It is hoped that this book will help persons take a step in the right direction when coming to grips with these issues. For some, this might be the guidance necessary to solve a problem for themselves. For others it might provide awareness that a problem exists and seek assistance from an expert. Uncertainty and variability are at best a nuisance and at worst a disaster. Even the most accomplished statistician cannot make uncertainties vanish. One can however be well served to recognize that they are present and make an effort to describe them quantitatively.

I would like to acknowledge the valuable assistance of Dr. James Durham who served as Administrative Dean for the Summer School. He provided the framework necessary to keep schedules on track. Julie Asmus in the Departmental office always managed to produce a smile and help as deadlines approached. Leilyn Perri served as the editor of special publications for the Health Physics Society. He did an admirable job of consolidating manuscripts that had many random deviations from the Guidelines for

Authors. Ryan Alexander helped me prepare illustrations that had the tendency to disappear somewhere in cyberspace just before completion. And finally, a big hug for Lulu.

Thomas B. Borak

CONTRIBUTORS

Thomas B. Borak
Department of Radiological Health Sciences
Colorado State University
200 West Lake Street
Ft. Collins, CO 80523
Tborak@cvmbs.colostate.edu

Phillip L. Chapman
Department of Statistics
Colorado State University
200 West Lake Street
Ft. Collins, CO 80523
Pchapman@stat.colostate.edu

Richard O. Gilbert
Battelle Washington Office
901 D Street, S.W., Suite 900
Washington, DC 20024
ro_gilbert@pnl.gov

Hari K. Iyer
Department of Statistics
Colorado State University
200 West Lake Street
Ft. Collins, CO 80523
(970) 491-5347
hari@stat.colostate.edu

Thomas B. Kirchner
Carlsbad Environmental Monitoring & Research Center
1400 University Drive
Carlsbad, NM 88220-3235
505-887-2759
kirchner@cemrc.nmsu.edu

Tom Tadfor Little
Los Alamos National Laboratory
P.O. Box 1663
Los Alamos, NM 87545
Tlittle@lanl.gov

Harry F. Martz
Group TSA-1, MS F600
Los Alamos National Laboratory
Los Alamos, NM 87545
(505) 667-2687
Fax (505) 667-4470
hfm@lanl.gov

Guthrie Miller
MS-E546, ESH-12
Los Alamos National Laboratory
P.O. Box 1663
Los Alamos, NM 87545
Guthrie@lanl.gov

James E. Turner
127 Windham Road
Oak Ridge, TN 37830
Jeturner@usit.net

John E. Wilson
Oak Ridge National Laboratory
Grand Junction Office
2597 B 3/4 Road
Grand Junction, CO 81503
wu6@ornl.gov

TABLE OF CONTENTS

Preface i

List of Contributors iii

Table of Contents iv

Chapter 1: **Concepts And Terminology For Probability And Statistics**
Thomas B. Borak 1

Chapter 2: **An Introduction to Bayes, Hierarchical Bayes, and Empirical Bayes Statistical Methods in Health Physics**
Harry F. Martz 55

Chapter 3: **An Introduction To Monte Carlo Methods**
James E. Turner 85

Chapter 4: **Uncertainty Analysis**
Thomas B. Kirchner 115

Chapter 5: **Applied Regression Methods**
Phillip L. Chapman 163

Chapter 6: **Nonparametric Statistical Methods**
Hari K. Iyer 213

Chapter 7: **Developing Field Sampling Plans Using The Data Quality Objectives Process**
Richard O. Gilbert and John E. Wilson 271

Chapter 8: **Bayesian Thinking**
Tom Tadfor Little 313

Chapter 9: **Use Of Bayesian Software Tools For Internal Dosimetry**
Guthrie Miller 323

Subject Index 331

Chapter 1

CONCEPTS AND TERMINOLOGY FOR PROBABILITY AND STATISTICS

Thomas B. Borak

1.1 INTRODUCTION

1.1.1 Objectives

This chapter is meant to serve as a brief survey or refresher course in preparation for the topics presented in the following chapters. It is by no means a complete summary of the entire vocabulary associated with probability and statistics. It should however provide the reader with sufficient information to proceed with the applications that are presented with more detail later in the book.

The chapter can be omitted by those already familiar with this material. There is a list of references consisting of books and monographs for the reader who wishes to obtain additional information or numerical examples relating to these introductory topics.

1.1.2 Fundamentals

Probability and statistics are two expressions that are closely associated with each other and often used interchangeably. Everybody recognizes that many observations in day-to-day life seem to be unpredictable. In some cases there is inherent variability. These are unavoidable variations such as how much rain will fall on Tuesdays, the weight of individuals or the stochastic nature of radioactivity. There are also uncertainties associated with making an observation. Even if the input is constant, results will not be known with complete precision because of measurement errors. Methods of probability and statistics are designed to help make sense out of this randomness in nature. However, applications of probability and statistics are distinctly different.

Probability begins with functions that are used to forecast random variations in the outcome of a selected process. Distributions of random results are computed so that we can deduce what we might expect to observe assuming that the functions are correct. Thus, probability is a form of deductive reasoning where known principles are used to form a logical set of observations.

In statistics one works in reverse. A sample of observations or data is used to infer what the underlying distribution must have been. Thus, statistics is a form of inductive reasoning where data are used to obtain parameters for functions that form general conclusions about the process.

This chapter will develop some of the basic framework of probability and statistics that will be used in the practical applications that are described in the following chapters.

1.2 PROBABILITY

1.2.1 Axioms of Probability

The results of an experiment or measurement are not known until the observation is completed. The set of all possible outcomes from an experiment is defined as S. One possible outcome is the result, A. The classical definition of probability is based on the chances or odds for observing a specific outcome A and is written as P[A]. This does not necessarily give information on other possible outcomes. It is simply the probability of observing the result described by A. Other more specific definitions contain statements on the frequency of observations as well as the subjective nature of interpreting a result. These will be discussed in some of the following chapters.

There are some fundamental properties associated with a probability, P[ ], that might seem trivial at first, but they form an important basis for applications in probability and statistics. It is interesting that the development of these axioms and relationships took almost 300 years from the time the Pascal and Fermat began thinking about these issues in reference to the art of gambling.

The first axiom is: $$P[\mathrm{A}] \geq 0 \tag{1.1}$$

This axiom says that the probability associated with the outcome A cannot be a negative number, but it could be zero. It establishes a lower bound for the numerical value of a probability.

The second axiom is: $$P[\mathrm{S}] = 1 \tag{1.2}$$

This axiom says that total probability for all possible outcomes in the set S is equal to 1. It establishes an upper bound for the value of a probability such that the probability for any single event cannot exceed 1.

Results that cannot occur simultaneously are defined to be mutually exclusive. This means that if you observe the event B you cannot observe A and vise versa. The third axiom states that if A and B are two possible outcomes that are mutually exclusive then:

$$P[\text{A or B}] = P[\text{A}] + P[\text{B}] \tag{1.3}$$

From these axioms it is possible to develop other important properties of probability. If $P[\text{A}^{\text{not}}]$ is the probability that A is specifically not observed or in other words, anything but A is observed, then:

$$P[\text{A}^{\text{not}}] = 1 - P[\text{A}] \tag{1.4}$$

The joint probability of observing two outcomes, A, B is P[A and B]. It is also written as P[A,B]. Two possible outcomes, A, B are said to be independent if:

$$P[\text{A and B}] = P[\text{A}] \bullet P[\text{B}] \tag{1.5}$$

There is also the concept of a conditional probability. $P[\text{A}|\text{B}]$ is the probability of observing A under the condition that B has already been observed. For example, what is the probability that a continuous air monitor will sound an alarm, P[A], if a ventilation fan happens to fail, P[B]. The relationship of this conditional probability to the joint probability of observing both A and B is

$$P[\text{A}|\text{B}] = \frac{P[\text{A and B}]}{P[\text{B}]} \tag{1.6}$$

For this to be valid, P[B] must be greater than zero. This relationship can be rearranged,

$$P[\text{A and B}] = P[\text{A}|\text{B}] \bullet P[\text{B}] \tag{1.7}$$

The joint probability can also be expressed the other way around:

$$P[\text{A and B}] = P[\text{B}|\text{A}] \bullet P[\text{A}] \tag{1.8}$$

These expressions can be combined such that,

$$P[\text{A}|\text{B}] = \frac{P[\text{B}|\text{A}] \bullet P[\text{A}]}{P[\text{B}]} \tag{1.9}$$

Relationships for probability may seem confusing at first glance but they form the basis for all modern analysis and it is worthwhile to become familiar with them.

1.2.2 Random Variables

There could be many results from an experiment or observation. The entire collection of possible outcomes forms a set of observables called S. The observables can be almost anything such as a name (P[Ford]), color (P[Blue]) or model (P[Explorer]). In many practical situations the result or observation is a number. A random variable, X, is defined as a transformation that converts or maps each outcome in the set of observables into a number having the value x. This value, x, is called a random number. The range for X may be limited to certain values or extend over the space of all numbers depending on the type of observation and set of outcomes. It is possible that several outcomes in S could yield the same the same random number x. For instance, when rolling two dice if X is the sum of both, there are several possibilities for getting the result the x = 7. The monthly dose registered by several personal dosimeters could be the same value, but the circumstances behind how the dose was received could be quite different.

1.2.3 Discrete Random Variables

In some situations a random variable can only produce random numbers that are discrete integers such as 0, 1, 2, 33, 455, etc. For example, if the random variable X is simply the outcome of the roll for a single die, the complete set of observables for X is a range of six integers 1,2,3,4,5,6. Another example of a discrete random variable would be results from a counting system with a detector connected to a scaler. When the scaler is on, counts are registered until the scaler is switched off. During this time the device might record zero counts, one count, two counts, but never one-half of a count, or five and one-third counts. Another result may occur if the observation is repeated, but it will always be an integer value.

1.2.4 Continuous Random Variables

Sometimes random variables can produce numbers that are not restricted to integers. In a counting system, the time between counts is variable without numerical restrictions. Another example would be the doses recorded on a sample of personal dosimeters. There may be a practical range of observed values but they would not necessarily be restricted to integers.

1.2.5 Probability Density Functions

The probability that a random variable X has the specific result A is written as $P[X=\mathrm{A}]$. It is often useful or even necessary to have an expression that gives a relationship of the probabilities for any value of $X \rightarrow \mathrm{x}$. That way one could immediately determine the probability for x or any range of values for x. This functional relationship is called the probability density function or PDF. It gives the structure or shape of the probability curve represented by the entire set of observables when the outcomes are mapped by the random variable, X, into the random number x.

The probability density function for the discrete random variable X is written as $f_X(\mathrm{x})$. It represents the probability that X takes on the value x:

$$f_X(\mathrm{x}) = P[X = \mathrm{x}] \tag{1.10}$$

An example of a discrete probability density function is shown in Fig. 1.1. Each bar in the histogram represents the probability for observing the value x. The bars in a sense resemble a mass or weight depending on the probability and thus the term probability density. The density functions $f_X(\mathrm{x})$ could have an unlimited variety of shapes and ranges. However, a PDF is normalized such that the total probability is one:

$$\sum_{\mathrm{x}=-\infty}^{\mathrm{x}=+\infty} f_X(\mathrm{x}) = 1 \tag{1.11}$$

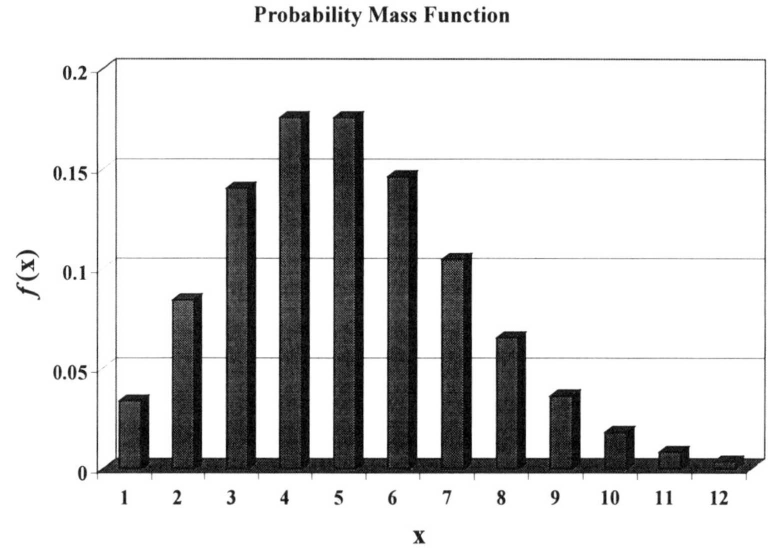

Fig. 1.1 A Probability Mass Function for discrete random variables.

For continuous random variables the probability density function is written in a form that is similar to that for discrete random variables. It must be emphasized that for continuous random variables, $f_X(\mathrm{x})$ is not the probability that the random variable takes on the value x. Probability for continuous random variables is analogous to the area under the curve represented by the PDF. The area under a single point is zero. Area must have height, $f_X(\mathrm{x})$ and width, dx. Thus, $f_X(\mathrm{x})$ should be considered as a continuous curve having the property that the probability for observing X in any interval between a and b is equal to,

$$P[\mathrm{a} \leq X \leq \mathrm{b}] = \int_{\mathrm{a}}^{\mathrm{b}} f_X(\mathrm{x})\mathrm{dx} \tag{1.12}$$

The density functions $f_X(\mathrm{x})$ for continuous random variables could have an unlimited variety of shapes and ranges. However, the PDF is normalized such that the total probability is one:

$$\int_{-\infty}^{+\infty} f_X(x)dx = 1 \qquad (1.13)$$

The median value of X for a PDF is defined as

$$\int_{-\infty}^{M} f_X(x)dx = 0.5 = \int_{M}^{\infty} f_X(x)dx \qquad (1.14)$$

Thus one half of all possible observations of X will be less than the value M and one half of the observations will be greater than M.

1.2.6 Cumulative Probability Distributions

Many times it is desirable to know the probability that the number of counts will be less than 100 or that the radioactivity in a sample will be less that 127 Bq. This process represents the accumulation of probabilities for the random variable of interest and is called the cumulative distribution function, CDF. For discrete random variables

$$\text{CDF} = P[\,X \leq \text{b}] = F_X(\text{b}) = \sum_{-\infty}^{\text{b}} f_X(x) \qquad (1.15)$$

and for continuous random variables,

$$\text{CDF} = P[\,X \leq \text{b}] = F_X(\text{b}) = \int_{-\infty}^{\text{b}} f_X(x)dx \qquad (1.16)$$

It represents that probability that the random variable X is less than or equal to b. In other words, $F_X(\text{b})$ is the probability that X will not exceed the value x. The CDF is a function that is never decreasing such that, for a < b, then $F_X(\text{a}) \leq F_X(\text{b})$. It has a lower bound of zero,

$$\lim_{\text{b} \to -\infty} F_X(\text{b}) = 0 \qquad (1.17)$$

and an upper bound of one,

$$\lim_{b \to +\infty} F_X(b) = 1 \tag{1.18}$$

This can be turned around if one wishes to know the probability that X will always be greater than the value b. This is referred to as the complementary cumulative distribution function, CCDF, and is defined for both discrete and continuous random variables as:

$$CF_X(b) = \int_b^{\infty} f_X(x)dx = 1 - F_X(b) \tag{1.19}$$

It is a function that is never increasing between the values of one and zero.

The relation ship between the PDF and CDF is

$$f_X(b) = \frac{dF_X(b)}{dx} \tag{1.20}$$

1.2.7 Quantiles

Very often it is important to focus on the extreme values of the random variable. These are referred to as the tails of the probability distribution. A quantile, x_γ can be used to identify the tails such that x_γ is the smallest number satisfying the following expression:

$$P[X \le x_\gamma] = \int_{-\infty}^{x_\gamma} f_X(x)dx = F_X(x_\gamma) = \gamma \tag{1.21}$$

Very often the quantile γ is expressed in terms of percent and is referred to as a percentile. Thus the 10% quantile, would be the random number $x_{0.1}$ such that the cumulative probability for observing values of X that are less than or equal to $x_{0.1}$ is 0.10

(10%). The 95% quantile would be the random number $x_{0.95}$ such that the cumulative probability for observing values less than or equal or equal to $x_{0.95}$ would be 0.95 (95%).

1.2.8 Expectation

A probability density function provides information on the odds or probability of obtaining any of the possible values for a random number. Occasionally one is interested only in a single question, "What value might I expect to observe from this distribution ?"

This can be generalized to any function or quantity that depends on a random variable, *G(X)*. Consider the case where there is a contaminated source of vegetables. The committed dose to an individual depends on the amount of vegetables eaten, which varies from person to person. Let *X* be the random variable representing the consumption of vegetables with the probability density function $f_X(x)$ and G(X) is the function that converts vegetable consumption, *X*, into committed dose *D*. Then the question might be, "What committed dose might I expect to obtain from this distribution of contaminated vegetables?".

For any function *G(X)* of the discrete random variable *X*, the expected value for this function is defined as:

$$E[G(X)] = \sum_{-\infty}^{+\infty} G(x)\cdot f_X(x) \tag{1.22}$$

For continuous random variables the expected value of the function is,

$$E[G(X)] = \int_{-\infty}^{+\infty} G(x)\, f_X(x)dx \tag{1.23}$$

Simply stated, the expected value of a function is the value for the function of a random number multiplied by the probability that the random number occurs and then summed over all possible values of the random number. There are certain relationships and definitions associated with probability distributions that are based on expectation.

If *G(X)* is a constant, A, then

$$E[A] = A \tag{1.24}$$

The expected value of the sum of two functions is.

$$E[G(X) + H(X)] = E[G(X)] + E[H(X)] \tag{1.25}$$

Perhaps the most frequently asked question is; "What is the expected value for the random number itself?" For this case G(X) = X and the expected value of for the continuous random variable X is,

$$E[X] = \int_{-\infty}^{+\infty} x\, f_X(x)dx = \mu \tag{1.26}$$

The integral is replaced with a summation for discrete random variables.

This expected value has particular significance. It is defined as the mean value of the probability distribution and has the symbol μ. Do not confuse this with the central value or most common observation. The mean is a weighted average and actually represents a balance point of the probability density function. Descriptively it is the value of X where the density function would balance if it were placed on a pivot point at the exact location X = μ.

The function $G(X)$ can become more complex For instance

$$E[X^2] = \int_{-\infty}^{+\infty} x^2 f_X(x)dx \tag{1.27}$$

represents the expected value of the square of the random variable X. This is not the square of the mean (μ^2) but could be called the mean of the square of the random variable. Taking the square root of this number is usually referred to as the Root Mean Square or RMS.

The last two expectations above are also referred to as moments (about zero) of the random variable X. The general definition of a moment is

$$\mu_k = E[X^k] \tag{1.28}$$

Thus the first moment μ_1 is identical to the $E[X]$ which is the mean, and the second moment is $E[X^2]$.

Another expectation of interest is

$$E[(X-\mu)^2] = \int_{-\infty}^{+\infty} (x-\mu)^2 f_X(x)dx = Var[X] \qquad (1.29)$$

This is the expected value of the square of the distance between the random number and the mean value. It is a measure of the variation of X around the mean and is defined as the variance, *Var*. Although this is the accepted representation for the spread of the distribution, the units of variance are in terms of X^2, which can be confusing when relating it to practical applications. For this reason the concept of standard deviation, σ, has been adopted to describe the spread of the probability density function and is defined as,

$$\sigma = \sqrt{Var} \qquad (1.30)$$

It can be shown that

$$Var[X] = E_X[(X-\mu)^2] = E[X^2] - \mu^2 \qquad (1.31)$$

1.2.9 Covariance

Eqn. 1.5 described a condition when two or more random variables are independent. In effect the joint probability is simply the product of the probabilities for each of the independent outcomes. There are situations when random variables are related to each other. For example, the probability that a release of radioactivity will arrive at a given location will depend on wind speed and direction. There may be meteorological conditions that create a relationship between the distributions of speed and corresponding direction. One measure of such a relationship between random variables is called the covariance. It is defined as

$$Cov[X,Y] = E[(X-\mu_x)\cdot(Y-\mu_Y)] \tag{1.32}$$

From Eqn. 1.31 it is obvious that that $Cov[X,X] = Var[X]$. If X and Y are independent then $Cov[X,Y] = 0$. It can be shown from Eqn. 1.32 that

$$Cov[X,Y] = E[X\cdot Y] - \mu_x\cdot\mu_Y \tag{1.33}$$

Very often if two random variables are not independent, they are said to be correlated. The correlation coefficient for this case is defined as

$$\rho(X,Y) = \frac{Cov[X\,Y]}{\sigma_X\,\sigma_Y} \tag{1.34}$$

The correlation coefficient can be positive or negative but is limited to $|\rho(X,Y)| \leq 1$.

1.2.10 Examples of Continuous Distributions

1.2.10.1 ***Uniform Distribution.*** The Uniform distribution represents the situation where the probability is constant for all values of X within a defined range. That is, all values of x are equally probable. Two parameters are required to define the uniform distribution. They are the maximum value b, and minimum value, a. The form of the probability density function is,

$$f_X(x) = \frac{1}{b - a} \tag{1.35}$$

$$\text{Median} = \frac{b + a}{2} \tag{1.36}$$

$$\text{Mean} = \frac{b + a}{2} \tag{1.37}$$

$$\text{Variance} = \frac{(b - a)^2}{12} \tag{1.38}$$

Fig. 1.2 shows the probability density function and cumulative distribution for a Uniform distribution.

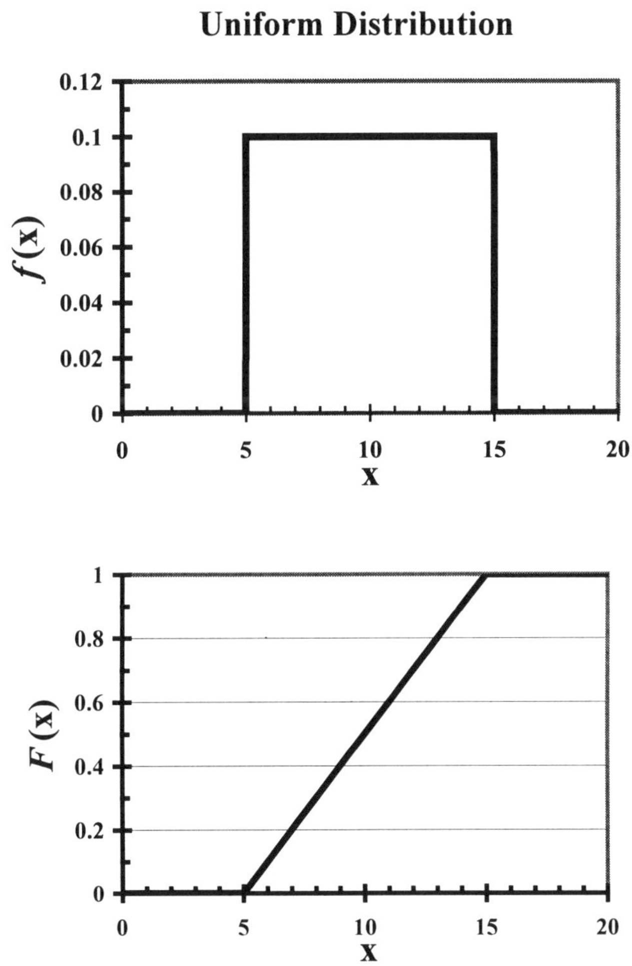

Fig. 1.2 The Probability Density Function (PDF) and Cumulative Distribution Function (CDF) for a Uniform distribution.

1.2.10.2 Triangular Distribution. The Triangular distribution rises linearly from zero to a peak value and then decreases linearly back to zero. Two parameters are required to define the Triangular distribution, m and k. It is defined over the range of values from minimum, X = m - k to maximum, X = m + k. The form of this probability density function is

$$f_X(x) = \frac{1}{k} - \frac{|x - m|}{k^2} \tag{1.39}$$

$$\text{Mean} = m \tag{1.40}$$

$$\text{Variance} = \frac{k^2}{6} \tag{1.41}$$

Fig. 1.3 shows the probability density function and cumulative distribution for a Triangular distribution. Eqn. 1.39 represents triangular distributions that are symmetric about the peak located at x = m. It is also possible to define triangular distributions that are not symmetric or have the shape of a right triangle.

1.2.10.3 Exponential Distribution. The exponential distribution can be defined for all positive values of X and by one parameter, λ that must be a number greater than zero. The form of this probability density function is.

$$f_X(x) = \lambda\, e^{-\lambda x} \tag{1.42}$$

$$\text{Median} = \frac{\ln 2}{\lambda} \tag{1.43}$$

$$\text{Mean} = \frac{1}{\lambda} \tag{1.44}$$

$$\text{Variance} = \frac{1}{\lambda^2} \tag{1.45}$$

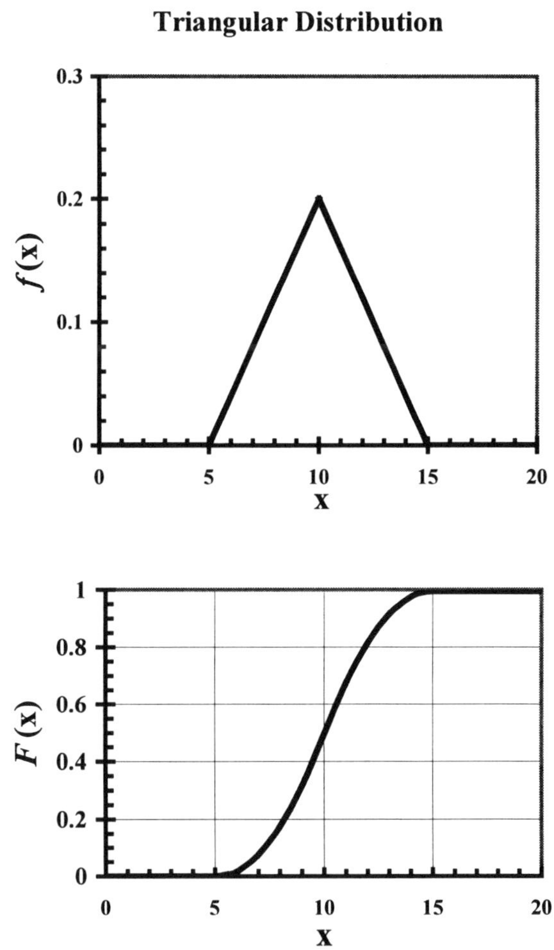

Fig. 1.3 The Probability Density Function (PDF) and Cumulative Distribution Function (CDF) for a Triangular distribution.

Fig. 1.4 shows the probability density function and cumulative distribution for an Exponential distribution. Note that the CDF for the Exponential distribution has the familiar form $F_X(x) = 1 - e^{-\lambda x}$.

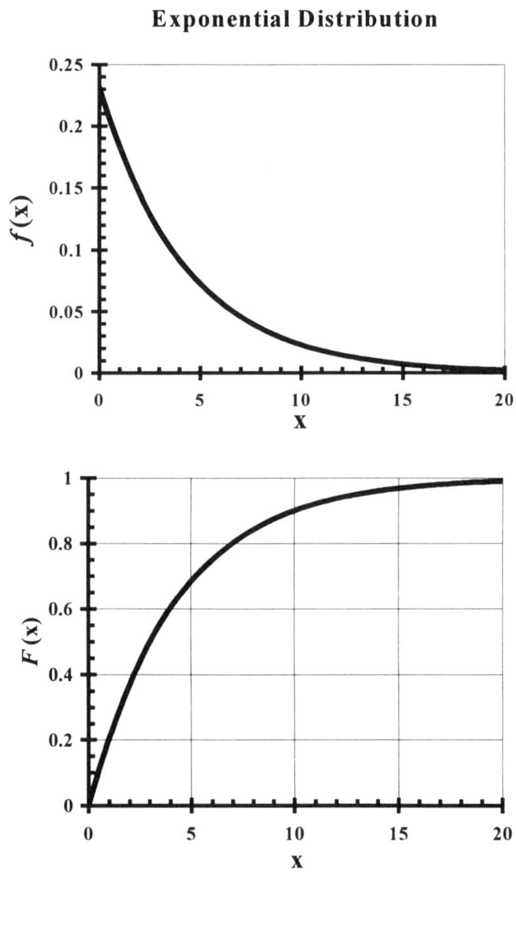

Fig. 1.4 The Probability Density Function (PDF) and Cumulative Distribution Function (CDF) for an Exponential distribution with parameter $\lambda = 3.0$.

1.2.10.4 Gamma Distribution. The Gamma distribution is defined for values of x that are greater than zero. Two parameters, α and β, are required to define the Gamma distribution. They must both be positive numbers and are sometimes referred to as the shape and scale parameters respectively. The form of this distribution is:

$$f_X(x) = \frac{\beta^{\alpha}\, x^{(\alpha-1)}\, e^{-\beta x}}{\Gamma(\alpha)} \qquad (1.46)$$

where $\Gamma(\ )$ is the Gamma function.

$$\text{Mean} = \frac{\alpha}{\beta} \qquad (1.47)$$

$$\text{Variance} = \frac{\alpha}{\beta^2} \qquad (1.48)$$

Fig. 1.5 shows the probability density function of Gamma distributions for several values of α when $\beta = 1$. Note that when $\alpha = 1$, the Gamma distribution reduces to an Exponential distribution with parameter $\lambda = \beta$.

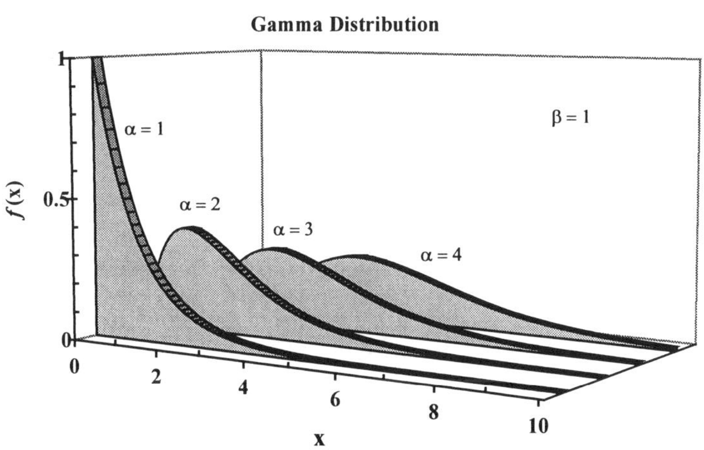

Fig. 1.5 The Probability Density Function (PDF) for a Gamma distribution with four values of the parameter α.

1.2.10.5 Beta Distribution The Beta distribution is defined for values of x that are greater than zero and less than one. Two parameters, a and b, are required to define the Beta distribution. They must both be positive numbers. The form of this distribution is:

$$f_X(x) = \frac{\Gamma(a+b)}{\Gamma(a)\,\Gamma(b)}\, x^{a-1}\,(1-x)^{b-1} \tag{1.49}$$

$$\text{Mean} = \frac{a}{a+b} \tag{1.50}$$

$$\text{Variance} = \frac{ab}{(a+b+1)(a+b)^2} \tag{1.51}$$

When a = b the beta distribution is symmetric about x = ½ and the variance decreases as the value of a increases. When a = b = 1 the beta distribution reduces to a Uniform distribution between 0 and 1. The distribution is skewed to the left when b > a and skewed to the right when a > b. Fig. 1.6 shows the probability density function of Beta distributions for several values of a and b.

1.2.10.6 Normal Distribution. The normal distribution can be defined for all positive and negative values of *X*. There are two parameters, μ and σ, required to define a Normal Distribution. The first parameter μ can be any real number and the second parameter, σ, must be a positive number. The form of this distribution is:

$$f_X(x) = \frac{1}{\sqrt{2\pi\sigma^2}}\, e^{-\left\{\frac{(x-\mu)^2}{2\sigma^2}\right\}} \tag{1.52}$$

$$\text{Mode} = \mu \tag{1.53}$$

$$\text{Mean} = \mu \tag{1.54}$$

$$\text{Variance} = \sigma^2 \tag{1.55}$$

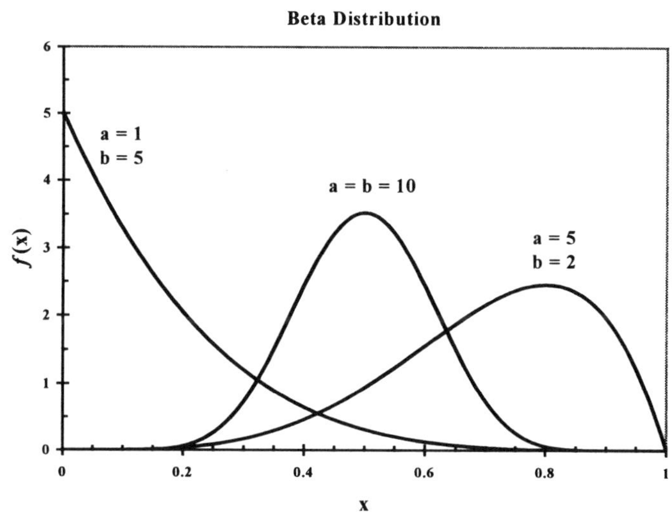

Fig. 1.6 The Probability Density Function (PDF) for a Beta distribution with several values of the parameters a and b.

Fig. 1.7 shows the probability density function and cumulative distribution for a Normal distribution. The distribution is perfectly symmetric with the mean and median having the same value. A special case of the Normal distribution is defined as the Standard Normal distribution when $\mu = 0$ and $\sigma = 1$. It has the form

$$f_Z(z) = \frac{1}{\sqrt{2\pi}} e^{-\frac{z^2}{2}} \tag{1.56}$$

There are several important properties that are unique to the Normal distribution. The probability that the random variable *X* occurs within an interval of σ surrounding the mean, $P[(\mu - \sigma) \le X \le (\mu + \sigma)]$, is 0.68. This is expressed using the probability density function as,

$$\int_{\mu - \sigma}^{\mu + \sigma} f_X(x)dx = 0.68 \tag{1.57}$$

The probability that the random variable X occurs within an interval of 2 σ surrounding the mean is $P[(\mu - 2\sigma) \le X \le (\mu + 2\sigma)] = 0.95$. Expressed in terms of the PDF this is:

$$\int_{\mu - 2\sigma}^{\mu + 2\sigma} f_X(x)dx = 0.95 \tag{1.58}$$

The interval about the mean that leaves a 5% probability in each tail is 1.65σ.

$$\int_{\mu - 1.65\sigma}^{\mu + 1.65\sigma} f_X(x)dx = 0.90 \tag{1.59}$$

1.2.10.7 Log-Normal Distribution. The Log-Normal distribution is defined as the probability distribution of a random variable *X* where the natural logarithm of *X*, is Normally distributed. It is defined only for values of *X* greater than zero. Two parameters, μ_g and σ_g, are required to define a Log-Normal distribution. Both must be positive numbers. The form of this distribution is

$$f_X(x) = \frac{1}{x\sqrt{2\pi}\,\ln(\sigma_g)}\, e^{-\frac{\{\ln(x)-\ln(\mu_g)\}^2}{2\{\ln\sigma_g\}^2}} \tag{1.60}$$

$$\text{Mode} = e^{(\ln(\mu_g) - \{\ln(\sigma_g)\}^2)} \tag{1.61}$$

$$\text{Mean} = e^{(\ln(\mu_g) + 0.5\{\ln(\sigma_g)\}^2)} \tag{1.62}$$

$$\text{Variance} = e^{[\{\ln(\mu_g)\}^2 + \{\ln(\sigma_g)\}^2]} \bullet (e^{\{\ln(\sigma_g)\}^2} - 1) \tag{1.63}$$

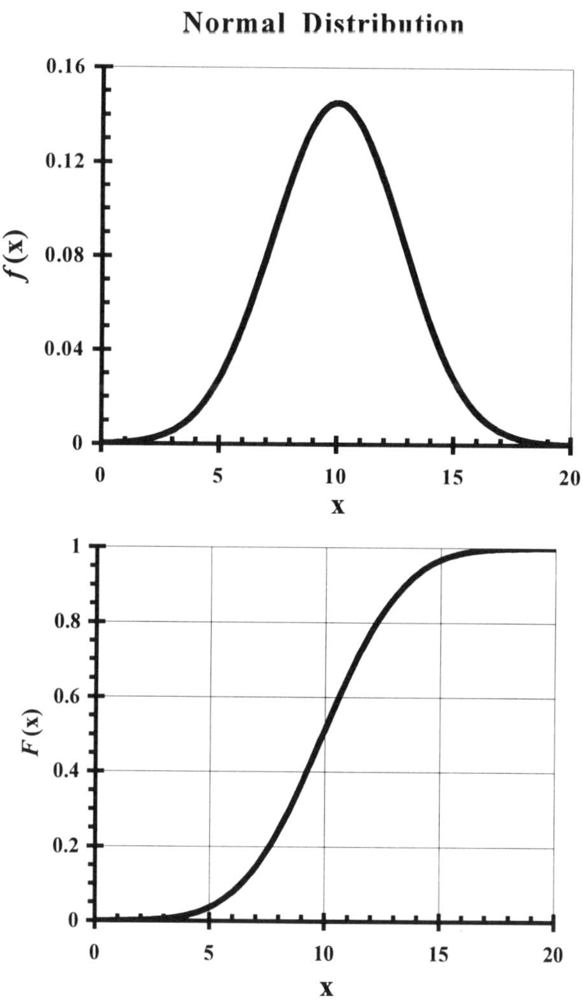

Fig. 1.7 The Probability Density Function (PDF) and Cumulative Distribution Function (CDF) for a Normal distribution.

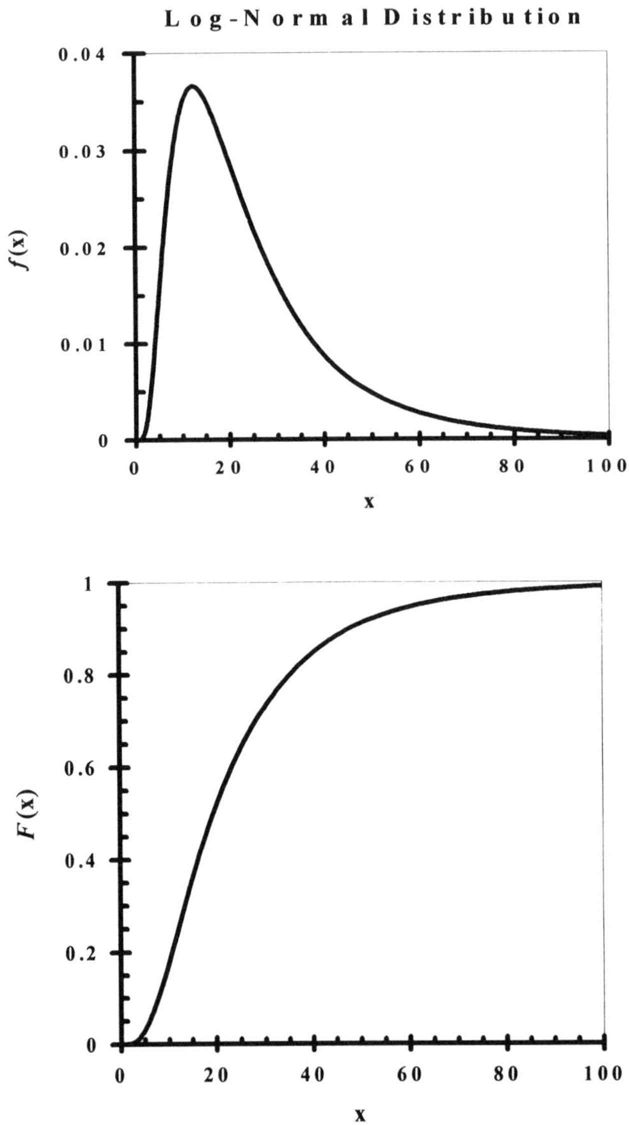

Fig. 1.8 The Probability Density Function (PDF) and Cumulative Distribution Function (CDF) for a Log-Normal distribution.

Fig. 1.8 shows the probability density function and cumulative distribution for the Log-Normal distribution. This distribution is characterized by a peak that decreases to 0 as x approaches zero and a long tail extending out to large values of x on the other end. The mode is equal to the median but the mean is always larger than the median.

As mentioned above, if the probability density function is plotted as $f_X(\ln x)d\ln x$ rather than $f_X(x)dx$, the graph would become symmetric and appear as a conventional normal distribution. The parameters μ_g and σ_g are often referred to as the geometric mean and geometric standard deviation. Since adding logarithms is equivalent to multiplying real numbers, the intervals of x corresponding to 68% and 95% probability are $P[(\mu_g \div \sigma_g) \leq X \leq (\mu_g \bullet \sigma_g)] = 0.68$ and $P[(\mu_g \div 2\sigma_g) \leq X \leq (\mu_g \bullet 2\sigma_g)] = 0.95$.

1.2.10.8 Chi-squared Distribution. The Chi-squared distribution (χ^2) is defined for all values of x that are greater than zero. There is one parameter, n, required to define the Chi-squared distribution. It must be an integer greater than zero and is referred to as the "degrees of freedom." The form of this distribution is

$$f_X(x) = \frac{\left[\frac{x}{2}\right]^{(n/2)-1} e^{-\frac{x}{2}}}{2\,\Gamma\left(\frac{n}{2}\right)} \tag{1.64}$$

$$\text{Mean} = n \tag{1.65}$$

$$\text{Variance} = 2n \tag{1.66}$$

Fig. 1.9 shows the probability density function and for several values of the parameter n. Note that when n = 2, the Chi-squared distribution reduces to an Exponential distribution with parameter $\lambda = \frac{1}{2}$.

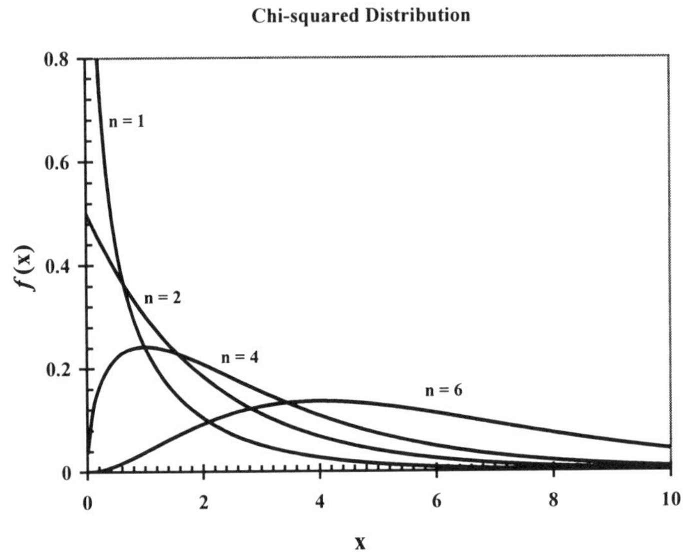

Fig. 1.9 The Probability Density Function (PDF) for a Chi-squared distribution with several values of the parameter n.

***1.2.10.9 Student t Distribution*.** The Student t Distribution is defined for all values of X that are greater than zero. There is one parameter, n, required to define the Student t distribution. It must be an integer greater than zero. The form of this distribution is

$$f_X(x) = \frac{\Gamma\left(\frac{n+1}{2}\right)}{\sqrt{n\,\pi}\;\Gamma\left(\frac{n}{2}\right)} \bullet \frac{1}{\left(1+\frac{x^2}{n}\right)^{(n+1)/2}} \tag{1.67}$$

$$\text{Mean} = 0 \tag{1.68}$$

$$\text{Variance} = \frac{n}{n-2} \quad \text{for } n>2 \tag{1.69}$$

Fig. 1.10 shows the probability density function when n = 1 and n = 5. It is always symmetric about zero. As the parameter $n \rightarrow \infty$, the Student t distribution becomes identical to the Standard Normal distribution. This is included for comparison in Fig. 1.10. When n = 1, the tails of the Student t distribution extend farther than the Standard

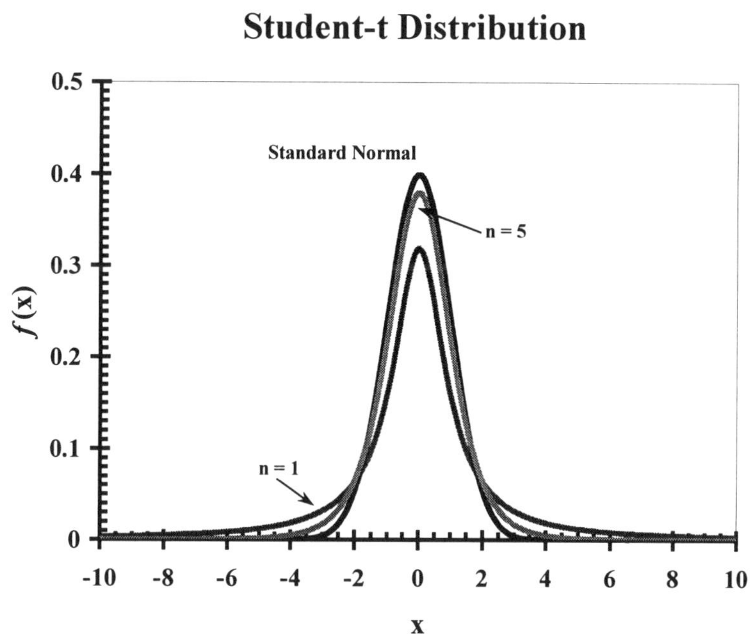

Fig. 1.10 The Probability Density Function (PDF) for a Student-t distribution with n = 1 and n = 5. Also shown is the PDF for a Standard Normal Distribution.

Normal distribution and this is reflected in the 95% quantile which is $t_{1;\ 95} = 6.31$ compared with $z_{95} = 1.65$ for the Standard Normal. One can see that the Student t distribution becomes quite similar to the Standard Normal distribution when n = 5.

1.2.10.10 Delta Function, The Delta function is a mathematical function that has the following properties:

$$\delta(x) = 0 \quad \text{when } x \neq 0 \tag{1.70}$$

and,

$$\int_{-\infty}^{+\infty} \delta(x)dx = 1 \tag{1.71}$$

To understand the nature of this function, imagine a peaked function something like a Normal distribution with a mean of zero and a standard deviation that approaches zero. As the standard deviation gets closer and closer to zero, the function must get higher and higher such that the area is always equal to 1. The exact shape of the function is not specified near x = 0, and for this reason it is classified as an improper function. However the integral must exist and equal unity.

The most important property of the δ function is:

$$\int_{-\infty}^{+\infty} \delta(x)f(x)dx = f(0) \tag{1.72}$$

where $f(x)$ is any continuous function. It is possible to shift the origin to create the delta function δ(x-a). This leads to the following relationship:

$$\int_{-\infty}^{+\infty} \delta(x-a)f(x)dx = f(a) \tag{1.73}$$

where a is any real number. Thus, the process of multiplying the function of x by δ(x-a) and integrating over all x is equivalent to solving the function at x = a.

Sometimes it is convenient to describe a hybrid probability density function as a linear combination of some continuous density function $g_X(x)$ and a delta function $\delta(x)$ such that

$$f_X(x) = \lambda g_X(x) + (1-\lambda)\delta(x) \tag{1.74}$$

This might be used to signify a situation corresponding to a sudden single intake of radioactivity at X = zero. If the mean and standard deviation of $g_X(x)$ is μ_0 and σ_0 respectively than it can be shown that

$$E[f_X(x)] = \lambda\mu_0 \tag{1.75}$$

$$Var[f_X(x)] = \lambda(\mu_0^2 + \sigma_0^2) - \lambda^2\mu_0^2 \tag{1.76}$$

1.2.11 Examples of Discrete Distributions

1.2.11.1 Bernoulli Distribution. There are many discrete probability distributions, but only a few have common applications in health physics. These are all derived from the simple concept of a Bernoulli distribution that can be described as individual trials each of which can only have two outcomes: success or failure. For example, when measuring radioactivity for a specified interval of time, each radionuclide in the sample could decay to be counted (success) or survive so that it cannot be counted (failure). Looking at a single atom, the probability that this atom will decay during the interval t will be called p. The probability that it will survive will be called q. Since these are the only two possibilities p + q = 1.

We then say that the random variable X has the value x = 1 when the trial is successful and the value x = 0 when the trial us unsuccessful. The probability density function is described as follows;

$$\begin{aligned} P[X = 1] &= p \\ P[X = 0] &= 1-p \end{aligned} \tag{1.77}$$

1.2.11.2 Binomial Distribution. A Bernoulli trial consists of a single atom having the probability of decaying p. This does not say that this specific atom will decay, but only that the odds of this happening are equal to p. The same argument can be made for every other atom in a sample being counted. The question is how many atoms will decay? To do this one needs to add up the number of successes. Using the Bernoulli representation, each atom that does decay (success) is assigned a 1 and each atom that survives (failure) is assigned a 0. The sum over all trials in the sample is the number of decays, *n*, which is a discrete number. If the process were repeated from the beginning, the number n might not be the same even though p for each atom has not changed. This illustrates the random nature of radioactivity. The objective is to understand the distribution of *n*.

If the sample of radioactivity originally contains N=10 atoms, then there will be a set of 10 Bernoulli trials. Since each one is independent, the probability for a given result is the product of all 10 probabilities. An example where 4 of the atoms decay would be:

$$p \cdot p \cdot p \cdot p \cdot q \cdot q \cdot q \cdot q \cdot q \cdot q = p^4 \cdot q^6$$

This is however not the only way this could happen. The following arrangement is also possible with different atoms decaying.

$$q \cdot q \cdot q \cdot q \cdot q \cdot q \cdot p \cdot p \cdot p \cdot p = p^4 \cdot q^6$$

Both cases have the same probability. The total probability for having 4 decays must consider all of the possible combinations.

The number of possible combinations for selecting n objects from a total population of N objects (without repetition) is:

$$\binom{N}{n} = \left(\frac{N!}{n!\,(N-n)!}\right) \tag{1.78}$$

For the example above where N=10 and n = 4, the number of possible combinations is

$$\binom{10}{4} = \left(\frac{10!}{4! \bullet 6!}\right) = 210$$

Therefore, the probability of observing 4 decays in the previous example would be

$P[n = 4] = 210 \bullet p^4 \bullet (1.p)^6$.

The general solution for the number of successes from a set of Bernoulli trials is called the Binomial Distribution. It has two parameters, N and p. N is the number of trials and is any integer greater than 1. The parameter, p, is the probability that any single trial will be a success and is any real number in the range 0 to 1. The form of the probability density function for the Binomial Distribution is,

$$f[n] = \binom{N}{n} p^n \bullet q^{N-n} \tag{1.79}$$

$$\text{Mean} = Np \tag{1.80}$$

$$\text{Variance} = Npq \tag{1.81}$$

This is an exact solution for the distribution of decays occurring during a timing interval where the probability that a single atom will decay is $p = 1\text{-}e^{-\lambda t}$ and $q = e^{-\lambda t}$ Here, λ is the radioactive decay constant.

1.2.11.3 Poisson Distribution. The Binomial distribution is correct for any N and any timing interval t. However, for many practical situations N can be very large. It becomes problematic to take factorials when N has values approaching Avogadro's number. In situations when the timing interval of interest is very small compared to the half-life of the radionuclide, p becomes very small. For instance, $p = 9.5 \times 10^{-7}$ when one is making a 15 minute count of ^{137}Cs which has a half-life of 30 y.

For the special case where N is very large and p is very small, but N•p is finite the Binomial Distribution can be approximated with the Poisson Distribution. The Poisson distribution has only one parameter, a, that can be any number greater than zero. For the example of radioactivity, a = Np. The form of the Poisson distribution is

$$f[n] = \frac{a^n e^{-a}}{n!} \tag{1.82}$$

$$\text{Mean} = a \tag{1.83}$$

$$\text{Variance} = a \tag{1.84}$$

A unique property of the Poisson Distribution is that the mean and variance are both numerically equal to the parameter a. Fig. 1.11 shows the probability density function for several values of the parameter a. The discrete random number, *n*, can only be an integer greater than or equal to zero. However, the mean of the distribution, a, can be any positive number.

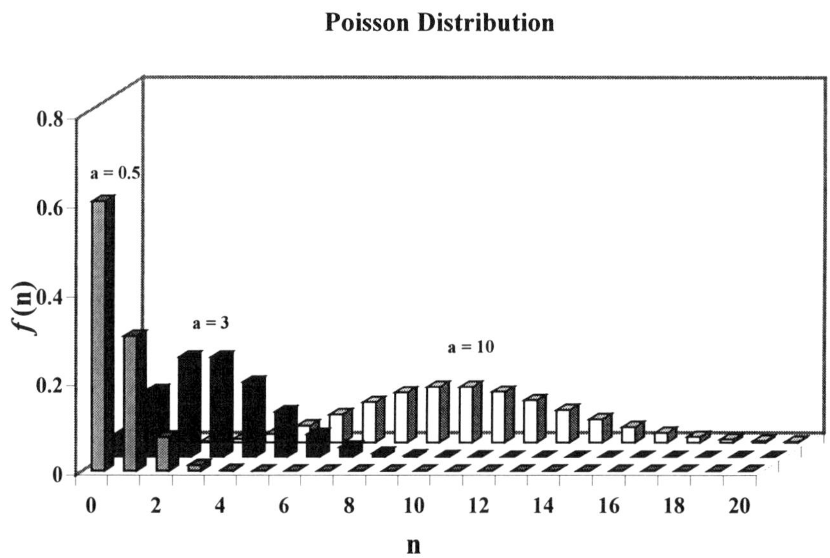

Fig. 1.11 The Probability Density Function (PDF) for a Poisson distribution with the parameter, a, having values of 0.5, 3 and 10.

The shape of the Poisson Distribution changes as the mean increases. When the mean becomes large (> 20) the distribution becomes symmetric. For this case the Poisson distribution can be approximated by a special discrete form of the Normal Distribution with the condition that the variance σ^2 is equal to the mean μ = Np. The probability that *n* will occur in an interval of σ surrounding the mean, $P[(\mu - \sqrt{\mu}) \leq X \leq (\mu + \sqrt{\mu})] = 0.68$, because for this special case $\sigma = \sqrt{\mu}$.

1.2.12 Central Limit Theorem

The central limit theorem is one of the most remarkable and useful results in probability theory. If $X_1, X_2, X_3, \ldots, X_n$ is a set of n independent random variables from the same distribution that has a mean μ and variance σ^2, then as $n \rightarrow \infty$ the following combination of these variables:

$$\frac{(X_1 + X_2 + X_3 + \ldots + X_n) - n\mu}{\sigma\sqrt{n}} \tag{1.85}$$

is distributed Normally with a mean of 0 and variance of 1. Specifically

$$\lim_{n \rightarrow \infty} P\left[a \leq \frac{(X_1 + X_2 + X_3 + \ldots + X_n) - n\mu}{\sigma\sqrt{n}} \leq b\right] = \frac{1}{\sqrt{2\pi}} \int_a^b e^{-\frac{x^2}{2}} dx \tag{1.86}$$

An equivalent expression can be made using the mean or average value of the sample. It states that $\bar{X}$ will have a Normal distribution with a mean μ and standard deviation $\sigma/\sqrt{n}$ as n becomes large. In less specific terms this theorem implies that the sum of a large number of independent random variables originating from the same distribution will be approximately Normal, regardless of what the original distribution looked like. Thus all of the properties of a normal distribution such as shape, confidence limits and hypothesis testing will apply to this sum. This is apparently why so many empirical observations in nature have the characteristic bell shape.

Thus for repeated sampling, the average value of the sample, $\bar{X}$, will follow a Normal distribution that has the same mean as the parent distribution. However the variance of the sample mean becomes smaller as the size of the sample increases. This value, $\sigma/\sqrt{n}$, is referred to the standard error of the mean. It is used in many applications that are described in following chapters.

1.2.13 Multivariate Probability Distributions

There are many situations when a result or process depends on two random variables. For instance we have seen that the number of decays occurring in a time interval is a random number n. A detector placed near the source will not record every event that actually occurs. Thus there is another random process which is the number of counts, C, registered by the counting system. What is the probability that n decays will occur and the detector will register C counts? This section describes the probability distributions when two random variables are involved.

1.2.13.1 Joint Probability Distributions. The joint probability density function for discrete random variables $f_{X,Y}(x,y)$ describes the simultaneous probability that the random variable X has the value x and the random variable Y has the value y.

$$f_{X,Y}(x,y) = P[\, X = x \text{ and } Y = y] \tag{1.87}$$

For continuous random variables the paired random variables X and Y are located on a plane defined by the x and y-axes. The joint density function describes the probability that the random variables X,Y, will lie in some region R of this x , y plane.

$$P[\,(X,Y) \text{ in R}] = \iint_R f_{X,Y}(x,y)dxdy \tag{1.88}$$

The probability $f_{X,Y}(x,y)$ associated with each point x,y in the X,Y plane can be imagined as a surface extending in the z direction. This is shown in Fig. 1.12. The normalization condition for a joint probability density function is

$$\int_{-\infty}^{+\infty}\int_{-\infty}^{+\infty} f_{X,Y}(x,y)dxdy = 1 \tag{1.89}$$

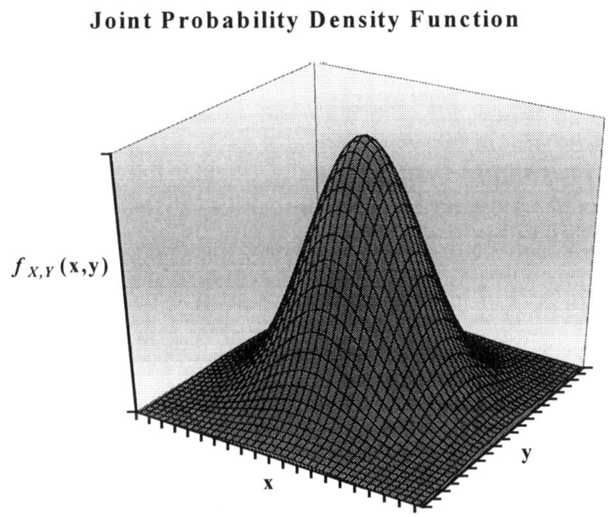

Fig. 1.12 An example of a Joint Probability Density Function (PDF) for two random variables *X* and *Y*.

1.2.13.2 Marginal Probability Distributions. Sometimes it is valuable to know the probability distribution for one random variable taking into account all values of the other random variables. For continuous random variables *X* and Y having the joint probability density $f_{X,Y}$(x,y,), the marginal density functions are:

$$f_X[x] = \int_{-\infty}^{+\infty} f_{X,Y}(x,y)dy \quad (1.90)$$

$$f_Y[y] = \int_{-\infty}^{+\infty} f_{X,Y}(x,y)dx \quad (1.91)$$

The marginal distributions can be explained by referring to Fig. 1.12. One marginal distribution would be the sum of the probabilities along the y-axis when X has the value of x. This is in effect summing or projecting the three dimensional surface onto either the x,z or y,z planes.

1.2.13.3 Conditional Probability Distributions. Consider the case where one of the two random variables is fixed. What is then the probability distribution of the other random variable? This is referred to as a conditional probability density function and is written as $f_{X,Y}[x \mid y]$. It is also possible to have it the other way around where $f_{X,Y}[y|x]$ is the probability for obtaining the value $Y = y$ given that $X = x$.

Although these two expressions look similar, they may be completely different distributions. The conditional distribution $f_{X,Y}[x \mid y]$ can be thought of as a slice of the three dimensional surface represented by $f_{X,Y}[x,y]$. This can be observed by referring back to Fig. 1.12. First one moves along the y-axis to the desired value y. At that point imagine viewing the shape of the probability density function in the x,z plane holding y fixed. Similarly $f_{X,Y}[y \mid x]$ could be viewed by moving along the x-axis to the desired value x, and viewing the density function in the y,x plane holding x fixed.

These are only a few of the more common probability distributions. In addition to the simple functional relationship for a PDF there can be modifications that are imposed depending on circumstances associated with the random process. Very often a distribution is used to describe a random variable but certain values for X are simply not realistic. For instance, even though the mathematical form of $f_X(x)$ exists for negative values of x, this is not realistic if X represents the mass of an object. For these cases the PDF might be truncated at $x = 0$. Truncation can be applied to the vertical axis also. It might be desirable to describe a system using a triangular distribution with a flat top. No matter what variation is applied to the PDF, the function must be adjusted to satisfy the normalization conditions described in Eqns. 1.11 and 1.13.

There are unlimited applications in health physics where one or more probability distributions can be used to forecast outcomes or model specific processes. Chapter 3 provides an introduction to the Monte Carlo techniques where random trials are simulated in succession and then combined to form a prediction of the outcomes. It is one way to determine the solution to an equation that would be difficult to obtain analytically. This is followed in Chapter 4 where uncertainties in individual components of a process are propagated together in order to obtain an estimate of the combined uncertainty using examples relating to radiation dose assessment.

1.3 STATISTICS

1.3.1 Estimation

Up until now we have focused on deductive reasoning where probability distributions are the basis for predicting outcomes. We now turn our efforts to applications where samples of data are used to infer the properties of the population that produced this data. This is known as statistical inference.

Probability distributions are expressed by the probability density functions $f_X(x)$. These are mathematical expressions that have one or more constants which are called parameters. For instance the Poisson distribution has one parameter, a, while the Normal distribution has two parameters, μ and σ. A general way of describing a PDF that has k parameters is:

$$f_X(x) = f_X(x\,;\theta_1, \theta_2, \ldots, \theta_k) \tag{1.92}$$

For simplicity the general designation for parameters will be the bold symbol $\boldsymbol{\theta}$ representing a vector that indicates there could be more than one parameter associated with the PDF $f_X(x\,;\theta)$.

When several measurements are made, the data are considered to be a sample of random numbers coming from a population described by $f_X(x\,;\theta)$. In this scheme the type of distribution, $f_X(\;)$, is known but the values for the parameters, θ, are unknown. It is assumed that this is all we know at this time and the objective is to estimate the parameters so that the entire population of events can be described.

A sample of data consists of n random numbers $x_1, x_2, \ldots, x_n$ that come from $f_X(x\,;\theta)$. We would like to form some combination or function of this data $W(x_1, x_2, \ldots, x_n)$ that can be used to estimate the true value one of the parameters θ. This function is called a statistic and the procedure is to find a statistic that is the best estimator for the true value of θ.

Estimation can be for a single number that is the best guess we can make from the data for the value of the unknown parameter. Estimation can also be for a range of numbers or an interval. The procedure is defined by specifying a specific probability, $P[\theta_{Low} < \theta < \theta_{High}] = p$ solving for θ_{Low} and θ_{High} based on the data such that if this interval were constructed for repeated data sets there would be a probability p that any particular interval would actually contain the true value of the parameter.

1.3.2 Point Estimation

Consider a function W that uses the data, x_1, x_2, ..., x_n to obtain an estimate for the value θ. If another set of data were to be used, W would most likely produce another value for θ. Thus W itself is a random variable that has it's own distribution. There are some criteria that are used to evaluate how good a point estimate may be.

A statistic W is considered to be unbiased if the expected value is

$$E[W] = \theta \tag{1.93}$$

Thus an unbiased estimator is one that on average will produce the correct value no matter how many data points are used in each trial. This is an extremely important criterion especially for estimators that are to be used repeatedly.

A statistic W is considered to be consistent if it converges to the correct value of θ as the sample size gets large.

$$\lim_{n \to \infty} (W) = \theta \tag{1.94}$$

There could be several functions (statistics) using the data that are unbiased estimators for a parameter. Since each statistic is also a random variable with a PDF, each has a variance about the mean (which is the true value since they are all unbiased). The best estimator W^* is the one such that

$$Var[W^*] \leq Var[W] \tag{1.95}$$

for all unbiased estimators W.

So far we have described only the properties of the estimators. An even more fundamental question is how does one find a function W? Many methods can be used to derive estimator functions for a specific PDF. Two of the most popular are the method of maximum likelihood and the method of moments.

1.3.2.1 Maximum Likelihood. Assume that $x_1, x_2, \ldots, x_n$ is a sample of independent random numbers from the probability density function $f_X(x\,;\,\theta)$. From Eqn. 1.5 the joint probability of getting this specific set of data is the product of the PDF evaluated at each value of x when the parameters have the value θ is:

$$P[x_1, x_2, \ldots, x_n \mid \theta\,] = f_X(x_1;\,\theta)\bullet f_X(x_2\,;\,\theta)\bullet\ldots\bullet f_X(x_n\,;\,\theta) \tag{1.96}$$

Now turn the problem around and consider that values of the data, $x_1, x_2, \ldots, x_n$, are fixed and define the joint probability above to be a function only of the unknown parameters θ. This is called the likelihood function and is written as

$$L(\theta) = P[\theta \mid x_1, x_2, \ldots, x_n] = \prod_{i=1}^{n} f_X(x_i\,;\,\theta) \tag{1.97}$$

The objective of maximum likelihood is to find the function W the maximizes $L(\theta)$. This is called the maximum likelihood estimate for θ. In most cases of interest, $L(\theta)$ is differentiable and finding the maximum is equivalent to solving for any one of the parameters θ when,

$$\frac{\partial L(\theta)}{\partial \theta} = 0 \tag{1.98}$$

Maximum likelihood estimators are always consistent and efficient for large sample sizes. However, in some cases they may be biased. Often the bias can be removed with a simple adjustment.

1.3.2.2 Method of Moments. Recall that the moments of a PDF are defined as

$$\mu_k = E[X^k] = \int_{-\infty}^{+\infty} x^k f_X(x\,;\,\theta)dx \tag{1.99}$$

Thus μ_k will be some function of the parameters θ that is a solution to the above integral.

If $x_1, x_2, \ldots, x_n$ is a random sample from the probability density function $f_X(x\,;\theta)$ with an unknown parameter, then an approximation to the moments for the PDF are the values:

$$m_k = \frac{1}{n}\sum_{i=1}^{n} x^k = \overline{x^k} \qquad (1.100)$$

The method of moments sets $\mu_k = m_k$. For example if there is only one parameter in the PDF, the statistic that is an estimator of θ is found by solving the equation for θ such that.

$$\mu_1 = m_1 = \bar{x}$$

If there are two parameters in the PDF, the statistics that form the best estimators are found by solving the simultaneous equations for each parameter:

$$\mu_1 = \bar{x}$$

$$\mu_2 = \overline{x^2}$$

1.3.2.3 Method of Least Squares. The method of moments and maximum likelihood were developed for a single random variable, X, and the assumption that the probability density function, $f_X(x)$, was known, but the parameters of the distribution were unknown. These methods do not apply if the PDF is unknown. There is another category of estimation that applies when one wishes to obtain the relationship between random variables. The most common situation is where the data consists of pairs of random numbers (X,Y). The question is how are the two related if the joint distribution is not known or how to interpret the conditional distributions such as $f_{Y|x}(y)$ versus x. In other words, how does the random variable Y depend on the random variable X.

For practical reasons this problem is usually simplified to that of asking what is the expected value of Y given x, $E[Y|x]$, rather than trying to obtain the complete description of $f_{Y|x}(y)$ versus x . Thus the objective is to determine the function $R(x)$ that is an acceptable representation of $E[Y|x]$. In many situations $R(x)$ is called a model. It can be shown that the following expectation:

$$E[(Y - R(X))^2] \tag{1.101}$$

is minimum when $R(X) = E[Y|x]$.

The process begins by selecting a suitable function and then using the data to estimate any parameters contained in that function. For example, the function might be assumed to be a polynomial

$$R(X) = r(x) = a_0 + a_1x + a_2x^2 + \ldots + a_kx^k \tag{1.102}$$

where the parameters a_0, a_1, …, a_k, need to be determined. The data consists of n pairs of points (x_1, y_1), (x_2, y_2), … , (x_n, y_n). The method of least squares uses the pairs of data to form the sum

$$L = \sum_{i=1}^{n} \{y_i - r(x_i)\}^2 \tag{1.103}$$

The best polynomial is one whose coefficients minimize L. If for example Y is assumed to be a simple linear function of X, then $y = r(x) = a + bx$. The coefficients a and b would be obtained by solving the following two equations

$$\frac{\partial L}{\partial a} = 0 \quad \text{and} \quad \frac{\partial L}{\partial b} = 0 \tag{1.104}$$

In general, the procedures for obtaining the relationship for $E[Y|x]$ is part of the general topic called regression analysis and is discussed in detail in chapter 5. In effect the methods allow you to quantitatively determine how much of the variation observed in the values of Y can be explained by the function or model $R(X)$. The model may or may not have a physical interpretation.

1.3.3 Interval Estimation

A test statistic, W uses the data $x_1, x_2, \ldots, x_n$ to obtain an estimate of a parameter θ. Interval estimation takes into consideration the fact that an estimate of a parameter θ using the statistic W is not unique because the value obtained will change when the data set changes. Thus there is a distribution of values for the estimate of θ. Although W, presumably provides the best point estimate for θ, what certainty is there that this estimate from one data set is the true value? The idea would be to express confidence for the estimate of a given parameter.

Since W produces a range of values for a given sample size there must be a PDF for the estimator $f_w(\theta)$. One idea would be to use this PDF to solve for the probability, p that θ lies between two specified numbers a and b:

$$P[\, a \leq \theta \leq b\,] = p \tag{1.105}$$

Another idea is the above in reverse which is to find the values a and b when the probability p is specified. If p = 0.9, then one can say that for repeated sampling, at least 90% of the intervals constructed this way will contain the true value of θ. Some examples of this process will be shown in the next section. By the way, there is nothing special or compelling with the selection of 90%. It could be 50% or 99%. Policy makers usually reach this decision, not statisticians.

1.3.4 Hypothesis Testing

Thus far we have described a process for estimating the parameters of a distribution that could either be point estimates or interval estimates. Statistical inference can also take another approach called testing. This process involves making assumptions about the distribution of the random variable such as implying that an unknown parameter θ has some specific value ($\theta = 1$) or range of values ($\theta \geq 1.5$). This is called a hypothesis H, and the objective is to test if the data supports or rejects this hypothesis. The result of hypothesis testing is a single decision "yes" or "no". In jurisprudence the analogy is "guilty" or "not guilty".

Formally there are two competing hypotheses called the null hypothesis, H_0 and an alternate hypothesis, H_1. In general the null hypothesis is assumed to be true until data convincingly proves this to be incorrect.

The process is defined as follows. A test statistic, $W(x_1, x_2, \ldots, x_n)$ represents a function of the observed data that will yield a value, w_n where n represents the number of data points in the sample. W is in effect a random variable that will yield different values of w_n for each set of data. Thus the test statistic will have a PDF written as $f_W(w_n)$. Once a test statistic and its PDF are selected a test must be identified that will be used to accept or reject the null hypothesis depending on values of w_n that are obtained from the data. For instance, the value $w_n = C$ could be a threshold that serves as a decision level for rejecting or accepting the null hypothesis. Thus you might accept H_0 when $w_n \leq C$ and reject H_0 when $w_n > C$. The range of values for w_n that reject the hypothesis is called the critical region. Identification of the test function W and determination of the critical region can be a very complex process. The trick is to find a function W that relates the data and the unknown parameter θ to the values w_n. However, when this is done, it is usually a simple matter of computing w_n from data at hand and comparing it with the critical region.

It is possible that the outcome of the decision may be incorrect. There are four possibilities associated with this as shown in Table 1. This is illustrated in Fig. 1.13. The curve is represents $f_W(w_n)$ when the null hypothesis is true. A decision level, C, is also shown. Whenever the data produces a value of $w_n \leq C$ the null hypothesis is accepted. The area under the curve for the range $w_n > C$ is the probability that the null hypothesis will be rejected even though it is true.

Table 1.1 - Two-way Decision Analysis

Decision	Null Hypothesis: True	Null Hypothesis: False
Reject H_0	Type I error (α)	Correct ($1-\beta$)
Accept H_0	Correct ($1-\alpha$)	Type II error (β)

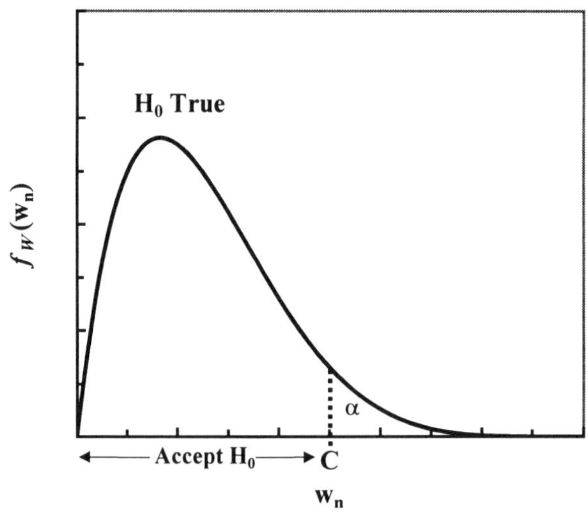

Fig. 1.13 The Probability Density Function (PDF) for a test statistic W_n when the null hypothesis, H_0 is true. C represents the critical value of w_n such that the probability of a Type I error, $P[w_n > C \mid H_0 \text{ true}] = \alpha$.

This is called a Type I error and is written as $P[w_n > C \mid H_0 \text{ True}] = \alpha$. The location of C depends on what is considered to be an acceptable value for α. How often are you prepared to make this mistake? To reduce the chances of falsely rejecting H_0, C should be selected so that α is as small as possible. The value, α, is sometimes called the significance level of the test.

There are problems with choosing α to be vanishingly small. This is shown in Fig. 1.14 where in addition to $f_W(w_n \mid H_0)$ for the null hypothesis, there is another curve $f_W(w_n \mid H_1)$ for an alternative hypothesis H_1. The second curve represents the probability that a data set will produce the value w_n if H_1 is true. Remember that you do not know apriori which hypothesis is true. However, whenever $w_n \leq C$ you will conclude that H_0 is true when in fact H_1 may be the correct hypothesis. If this occurs it is called a Type II error and is written as $P[w_n \leq C \mid H_1 \text{ True}] = \beta$. Type II errors are always coupled with both H_1 and H_0. You will notice that if α for H_0 is reduced by increasing the decision level, C, the value of β for H_1 will increase.

If β is the probability of accepting H_0 when H_1 is true, then $1-\beta$ is the probability of rejecting H_0 when H_1 is true. This is called the power of the test and is a measure of the ability to correctly conclude that H_0 is false. Very often two or more tests can be developed for testing the same hypotheses. All other things being equal, the best test is the one with the highest power.

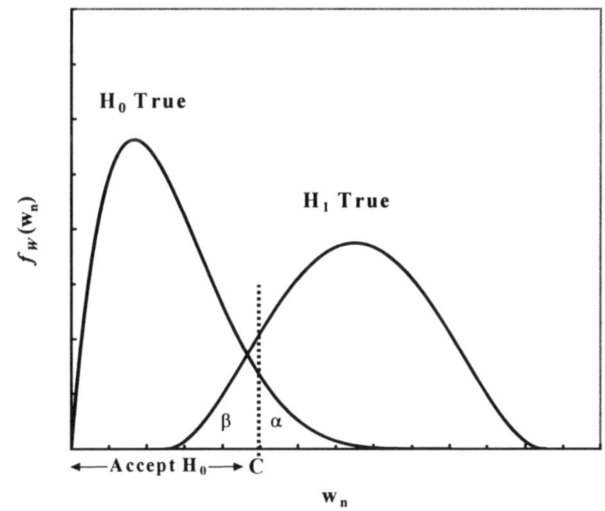

Fig. 1.14 Two Probability Density Functions for a test statistic W_n. One represents the case when the null hypothesis, H_0, is true and the other when the alternate hypotheses,H_1, is true. C represents the critical value of w_n such that the probability of a Type I error, $P[w_n > C \mid H_0 \text{ true}] = \alpha$ and the resulting Type II error $P[w_n < C \mid H_1 \text{ true}] = \beta$.

Often the result of a statistical test is reported using a P-value rather than the significance level α. This is called a significance probability. It is the smallest value of α for which the null hypothesis can be rejected based on the evidence contained in the data. Quoting a P-value is a superior way to report the results of a statistical test. It measures the amount of evidence in the data to support the alternative hypothesis. The P-value approaches zero as evidence in favor of the alternative hypothesis increases.

1.3.5 Estimation for Normal Distributions

There are some special results that are frequently applied when the random variables are Normally distributed. For this case the data consists of a sample of n independent random variables, $X_1, X_2, \ldots, X_n$ from the parent distribution. The objective is to use this data to answer questions about the two parameters μ, and σ.

The parameter, μ, for the Normal distribution happens to be identical to the mean. Thus the maximum likelihood estimator for μ is:

$$\hat{\mu} = \frac{1}{n} \sum_{i=1}^{n} x_i = \bar{x} \qquad (1.106)$$

Likewise, the square of the parameter σ happens to be identical to the variance so that the maximum likelihood estimator is:

$$\hat{\sigma}^2 = \frac{1}{n} \sum_{i=1}^{n} (x_i - \bar{x})^2 \qquad (1.107)$$

The estimate $\hat{\mu} = \bar{x}$ is an unbiased and consistent estimator for μ. However, $\hat{\sigma}^2$ is a biased estimator for σ^2. Another estimate for σ^2 that is both unbiased and consistent is called the sample variance S^2:

$$S^2 = \frac{1}{n-1} \sum_{i=1}^{n} (x_i - \bar{x})^2 \qquad (1.108)$$

Each data set $X_1, X_2, \ldots, X_n$ will provide different values for $\bar{X}$ and S^2. Thus they are also random variables. What testing can be done with specific results $\bar{X} = \bar{x}$ and $S^2 = s^2$ to estimate values or confidence limits for the two parameters, μ and σ?

It can be shown that if $X_1, X_2, \ldots, X_n$ are distributed normally with parameters μ and σ then $\bar{X}$ is also distributed Normally with a mean μ and variance σ^2/n. At first glance it looks like the problem is already solved. Just use the confidence intervals for a Normal distribution for $\bar{X}$ such as $P[\,\bar{x} \pm s/\sqrt{n}\,] = 0.68$. Unfortunately this does not work because there is no direct way to solve specifically for μ or σ from this relationship. A more careful approach is required.

***1.3.5.1 Estimating** σ.* We first begin with estimating the parameter σ. The sample variance S^2 is not Normally distributed. However the following test statistic:

$$Q = \frac{(n-1)\,S^2}{\sigma^2} \tag{1.109}$$

is a random variable that is distributed as Chi-squared with the parameter n-1. This is quite fortunate since Q depends on n and S^2 that we can get from the data and only one of the unknown parameters, σ. Also, Q is a random variable that has a known distribution with only one parameter, n-1, that is trivially obtained from the data.

For example, consider data form a normal distribution with a sample size of n = 11. For this case, Q is distributed according to χ^2_{10} as shown in Fig. 1.15. This distribution is not symmetrical. The locations of the 5% and 95% quantiles are also shown in Fig. 1.15. Thus 90% of the time of Q will occur between $\chi^2_{10;\,.05} = 3.94$ and $\chi^2_{10;\,.95} = 18.3$. For any sized sample, Eq. 1.102 can be rearranged to solve for σ^2 in terms of n, S^2 and Q. By substituting the 5% and 95% percentiles of χ^2_{10} for Q one can say that repeated sampling intervals defined by:

$$\left[\frac{(n\text{-}1)\, s^2}{\chi^2_{10;\, .95}} \, , \, \frac{(n\text{-}1)\, s^2}{\chi^2_{10;\, .05}} \right] \tag{1.110}$$

will contain the true value of σ^2 90% of the time.

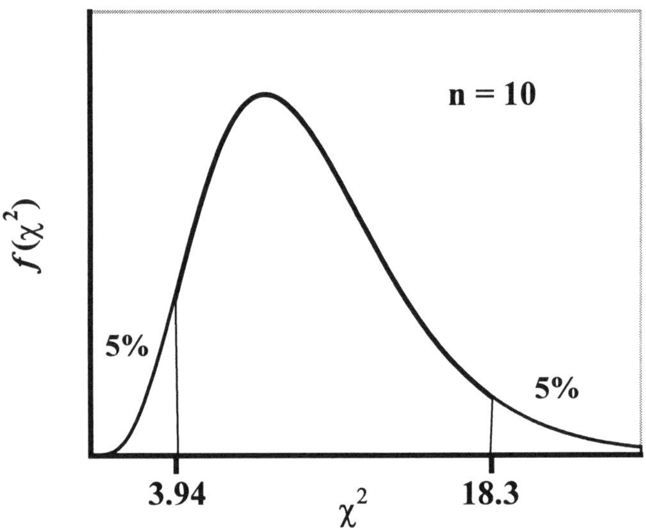

Fig. 1.15 The Probability Density Function for a Chi-squared distribution with the number of degrees of freedom, n, equal to 10. Also shown are locations of the 5% and 95% quantiles.

This approach can also be used for hypothesis testing. Consider that you want to test the hypothesis that the true value of the unknown parameter is a specific value σ_0^2. Then

$$H_0: \sigma^2 = \sigma_0^2$$

$$H_1: \sigma^2 \neq \sigma_0^2$$

From the results above you reject the hypothesis H_0 with 90% confidence if:

$$\left\{\frac{(n-1)\,s^2}{\sigma_0^2}\right\} < \chi^2_{n-1;\,.05} \quad \text{or} \quad \left\{\frac{(n-1)\,s^2}{\sigma_0^2}\right\} > \chi^2_{n-1;\,.95} \tag{1.111}$$

1.3.5.2 ***Estimating*** μ. We now turn to the problem of setting confidence limits for the unknown parameter μ. There is another remarkable relationship for random variables that are Normally distributed. If $X_1, X_2, \ldots, X_n$ are independent and normally distributed with parameters μ and σ, then a test statistic T can be is defined as:

$$T = \frac{\overline{X} - \mu}{\frac{S}{\sqrt{n}}} \tag{1.112}$$

The random variable T happens to be distributed as a Student t distribution with the parameter n-1. Once again we have the good fortune of having a random variable T that

depends on n, $\overline{X}$ and S^2 which we can get from the data and only one of the unknown parameters. Also, T is a random variable that has a known distribution with only one parameter, n-1, that is also trivially obtained from the data. From the Student t distribution there is 90% probability that T will be between the values $t_{n-1;\,.05}$ and $t_{n-1;\,.95}$. As shown earlier, the Student t distribution is always symmetric with a median value of zero. Because of the symmetry about zero

$$t_{n-1;\,.05} = -t_{n-1;\,.95} \tag{1.113}$$

Rearranging Eqn. 1.105 to solve for μ and inserting the 5% and 95% quantiles, one can say that that for repeated sampling, intervals defined by:

$$\left[\bar{x} - t_{n-1;.95} \bullet \frac{s}{\sqrt{n}} \quad , \quad \bar{x} + t_{n-1;.95} \bullet \frac{s}{\sqrt{n}} \right] \tag{1.114}$$

will contain the true value of μ 90% of the time.

This approach can also be used for hypothesis testing. Consider that you want to test the hypothesis that the true value of the unknown parameter is the specific value μ_0. Then

$$H_0: \mu = \mu_0$$

$$H_1: \mu \neq \mu_0$$

From the results above you reject the hypotheses with 90% confidence if H_0 if

$$\left\{ \frac{\bar{x} - \mu_0}{\frac{s}{\sqrt{n}}} \right\} < -\, t_{n-1;.95} \quad \text{or} \quad \left\{ \frac{\bar{x} - \mu_0}{\frac{s}{\sqrt{n}}} \right\} > t_{n-1;.95} \tag{1.115}$$

1.3.5.3 Comparing Two Sets of Data. The previous discussion involved estimation and hypothesis testing when only one data set was involved. Another common problem is based on deciding if two samples of data originated from the same or different distributions. Consider the case where one sample of data consists of $X_1, X_2, \ldots, X_n$ that are each normally distributed random variables with parameters μ_X and σ . Another sample of data consists of $Y_1, Y_2, \ldots, Y_m$ that are each normally distributed random variables with parameters μ_Y and σ^2. Thus both samples are derived from Normal distributions that have the same variance but different means. The pooled variance of the samples is defined as:

$$S_p^2 = \frac{(n-1)S_X^2 + (m-1)S_Y^2}{(n+m)-2} \tag{1.116}$$

The two sets of data can be combined to form another random variable $\tilde{T}$ such that

$$\tilde{T} = \frac{(\bar{X}-\bar{Y}) - (\mu_X - \mu_Y)}{S_p \bullet \sqrt{\frac{1}{n} + \frac{1}{m}}} \quad (1.117)$$

It can be shown that $\tilde{T}$ is a random variable that is distributed as a Student t distribution with the parameter n+m-2. Thus $\tilde{T}$ depends on $\bar{X}$, $\bar{Y}$, n, m, and S_p that we can compute from the data and $(\mu_X - \mu_Y)$ which we would like to determine. Also, $\tilde{T}$ is a random variable that has a known distribution with only one parameter, n+m-1, that is also trivially obtained from the data.

First one can find a confidence interval for the difference between the means $(\mu_X - \mu_Y)$. From the Student t distribution there is 90% probability that $\tilde{T}$will be between $t_{n+m-2;\,.05}$ and $t_{n+m-2;\,.95}$. Since the distribution is symmetric about zero, $t_{n+m-2;\,.05} = -t_{n+m-2;\,.95}$. Using these values and solving for $(\mu_X - \mu_Y)$ in Eqn. 1.109 one can say that for repeated sampling, intervals defined by:

$$\left[\{\bar{x}-\bar{y}\} - t_{n+m-2;\,.95} \bullet s_p \bullet \sqrt{\frac{1}{n} + \frac{1}{m}} \;,\; \{\bar{x}-\bar{y}\} + t_{n+m-2;\,.95} \bullet s_p \bullet \sqrt{\frac{1}{n} + \frac{1}{m}}\right] \quad (1.118)$$

will contain the true value for the difference between the means, $(\mu_X - \mu_Y)$, 90% of the time.

This approach can also be used for hypothesis testing. Consider that you want to test the hypothesis that both samples of data actually came from the same distribution. Thus the two parameters μ_X and μ_Y must be identical. Then:

$$H_0: (\mu_X - \mu_Y) = 0$$

$$H_1: (\mu_X - \mu_Y) \neq 0$$

The null hypothesis would be rejected with 90% confidence if :

$$\frac{(\bar{x}-\bar{y}) - (\mu_X - \mu_Y)}{s_p \bullet \sqrt{\frac{1}{n} + \frac{1}{m}}} < -t_{n+m-2;\ .95} \tag{1.119}$$

$$\frac{(\bar{x}-\bar{y}) - (\mu_X - \mu_Y)}{s_p \bullet \sqrt{\frac{1}{n} + \frac{1}{m}}} > t_{n+m-2;\ .95} \tag{1.120}$$

Section 1.3.5 has provided some samples of estimation and hypothesis testing when the data are derived from a Normal distribution. They were in the form of analytical solutions to the problem that reduced to equations where one only needs to substitute values derived from the data. There are similar approaches when the data are derived from other distributions. The methods are similar but the test statistic may follow something other than a Student t or Chi-squared distribution. In order for these methods to be valid the experimenter must insure that the data satisfies the stated assumptions for the underlying distributions and conditions for the associated parameters. Validating these assumptions is sometimes more challenging that estimating the unknown parameters. In this sense some subjective intuition often creeps into the analysis.

1.3.5.4 Analysis of variance. A natural extension of the two-sample tests described in section 1.3.5.3 is the case where the data is obtained from many sets or conditions. An example would be the distribution coefficient, K_d , of Pu in soil. Data for a particular type of soil will have variations. The objective is to determine if the estimate of K_d for a specific soil is different from other soils or just a reflection of the variance observed within the samples for each type of soil. The appropriate test statistic is the ratio of the estimated variances and the procedures developed for the estimation process is called analysis of variance (ANOVA). Similar methods are often applied as prospective tools that form the basis for experimental design. The objective is to plan the experiment in order to reduce the variance and improve the chances of rejecting a false hypothesis.

1.3.5.5 Non parametric analysis. Occasionally the data does not conform to any of the conventional probability distributions. For these cases, the underlying PDF responsible for the data may not be a simple expression of a random variable and some parameters. Yet there exists the need to test and compare hypotheses. Methods of statistical inference that do not rely on any specific probability distribution are often referred to as nonparametric analyses. These will be described in more detail in chapters 6 and 7 of this book.

1.3.6 Bayesian Inference

The inference methods described above began with the assumption that the form of the underlying distribution was known, $f_X(x;\theta)$, but values for the parameters of the distribution, θ, were unknown. Here x could be considered as the data and θ related to the cause for the data. The data were then used to estimate values or confidence intervals for these parameters. The only available information was the data at hand. No reference was made to previous results or experience. Thus this may be considered as a pure objective approach to statistical inference.

Another approach is called Bayesean inference. In this case parameters that are responsible for the observed data, Θ, are considered to have probability distributions of their own that represent our best state-of-knowledge or belief about Θ prior to consideration of the data x. The method combines information contained in the current data with information that might be available from previous or prior experience which may include subjective judgments. By incorporating knowledge obtained from prior experience, it becomes possible to infer values for the parameters from the data, rather than deducing probable values from the data given some assumed parameters. These methods were originally proposed by the mathematician Thomas Bayes. The approach is based on Bayes formula.

Suppose that the set of observables, S, is made up of mutually exclusive events X_i such that every outcome in S is represented by one and only one X_i. This might be the number of counts observed in a scaler for a specific counting interval. Thus X_i could only be one of the following results: 0,1,2,3,...,N where N is the total number of atoms in the sample being analyzed. Also suppose that S depends on another parameter, Θ that is related to the cause for the observed counts. This might be true radioactivity in a sample that can vary from sample to sample. Then for a specific event X = X

$$P[X] = \sum_{i=1}^{m} P[X|\Theta_i] \bullet P[\Theta_i] \tag{1.121}$$

For the example, this states that the probability of observing X counts in the detector is the probability of getting X counts in the detector when the true activity is Θ_i , $P[X|\Theta_i]$, times the probability that Θ_i is the true activity, $P[\Theta_i]$. This is then summed over all possible values of Θ_i. Here the true activity is assumed to have discrete values.

From Eqn.1.7 and 1.8 the joint probability of the event $P[X,\Theta_j]$ can be rearranged in terms of conditional probabilities such that

$$P[\Theta_j|X] = \frac{P[X|\Theta_j] \bullet P[\Theta_j]}{P[X]} \tag{1.122}$$

But if all of the observables X are mutually exclusive, then from Eqn. 1.114

$$P[\Theta_j|X] = \frac{P[X|\Theta_j] \bullet P[\Theta_j]}{\sum_{i=1}^{n} P[X|\Theta_i] \bullet P[\Theta_i]} \tag{1.123}$$

This is referred to as Bayes theorem. The formula permits one to reverse the direction of the conditioning and thus evaluate certain kinds of inverse probability. If one knows $P[X|\Theta_j]$ for every j , then one can use these to compute probabilities in the other direction $P[\Theta_j|X]$.

Using the example discussed above, let C be the number of counts registered by the detector and A be the true activity in the sample. Then $P[C|A]$ is the probability of getting C counts when the activity is exactly A. For example, if a specific counting interval T is very small compared to the half-life of the radioisotope, the distribution of counts registered by the detector can be described by the following Poisson distribution:

$$P[C|A] = \frac{e^{-AT\varepsilon} (AT\varepsilon)^C}{C!} \tag{1.124}$$

For this case ε is the counting efficiency. However, we are very often confronted with the reverse question. What is the activity in the sample when the detector measures a specific number of counts, $P[A|C]$? This can be obtained using Bayes formula from Eqn. 1.123

$$P[A|C] = \frac{P[C|A] \bullet P[A]}{\sum_{A=0}^{m} P[C|A] \bullet P[A]} \tag{1.125}$$

Using the nomenclature associated with Bayesian inference, *P*[C|A] is referred to as the likelihood function and must be known. It is the conditional response that the outcome C will occur for specific values for A, T and ε.

P[A] is referred to as the prior distribution. It represents the distribution of possible outcomes for A. This is the subjective part of Bayesian inference. Generally *P*[A] is described by a probability density function $\pi(A|\Theta)$ where Θ are the parameters for the prior distribution. The exact values for these parameters are unknown and may have distributions of their own. They must be determined without using the current measurement of C. The type of distribution $\pi(\ )$ and the properties of the parameters Θ are obtained from any available prior information including previous experience, expert opinion and common sense.

The denominator in Eqn. 1.125 represents the partitioning of *P*[C] and technically serves the role of a normalizing factor.

P[A|C] is called the posterior distribution that tells us the probability that A occurred based on both the likelihood function using the current observation and a prior distribution. This is an appealing feature of Bayesian inference since the result is obtained from new information (the data) as well as knowledge from previous experience (the prior distribution).

Chapter 2 presents a more detailed introduction to Bayesian methods. Chapter 8 is a discussion relating to some of the thought processes associated with Bayesian inference and chapter 9 contains examples of analysis using bioassay data.

1.4 REFERENCES

Box, G.E.P., Hunter, W.G. and Hunter, J.S. Statistics for experimenters. New York: John Wiley and Sons; 1978.

Brownlee, K.A., Statistical theory and methodology in science and engineering. Second Edition. New York: John Wiley and Sons; 1965

Chambers, J.M.: Cleveland, W.S.; Kleiner, B. and Tukey, P.A., Graphical methods for data analysis. Belmont,Ca: Wadsworth International;1983.

Larson, R.J. and Morris, L.M. An introduction to mathematical statistics and its applications. Second edition. Englewood Cliffs, NJ: Prentice-Hall, 1986.

Manley, B.F.J., Randomization bootatrap and monte carlo methods in biology. Second edition. London :Chapman and Hall: 1997.

Mood, A.M.; Graybill, F.A.; Boes, D.C. Introduction to the theory of statistics. Third edition. New York:McGraw-Hill.1974.

Ott, R.L. An Introduction to statistical methods and data analysis. Fourth edition.Belmont,CA: Duxbury Press:1993.

Ross, S. A first course in probability. Third edition, New York: McMillan Publishing Company; 1984.

1.5 AUTHOR PROFILE

Dr. Thomas B. Borak received a B.S. degree in physics from St. John's University in Collegeville, Minnesota and a Ph.D. degree in high-energy particle physics from Vanderbilt University in Nashville, Tennessee. He worked in radiation physics at the Fermi National Accelerator Laboratory, CERN, and Argonne National Laboratory. Presently Dr. Borak is a Professor in the Department of Radiological Health Sciences at Colorado State University and is Certified by the American Board of Health Physics with over 50 publications in peer reviewed literature. He has served as a consultant to the Governor of Colorado and as a member of committees for the National Academy of Sciences and the National Council on Radiation Protection and Measurements.

Chapter 2

AN INTRODUCTION TO BAYES, HIERARCHICAL BAYES, AND EMPIRICAL BAYES STATISTICAL METHODS IN HEALTH PHYSICS

Harry F. Martz

2.1 BAYESIAN STATISTICAL INFERENCE

As illustrated by the recent literature, Bayesian methods are widely and increasingly being used in scientific research. For example, a recent sampling of one popular physical and engineering sciences electronic database indicates Bayesian applications in such diverse areas as psychometrics, dairy science and agronomy, hydrology and hydraulics, environmetrics and ecology, planetary and space science, signal and image processing, econometrics, astronomy, condensed matter physics, molecular biology, expert systems, data mining and classification, epidemiology, quality and reliability, health physics, medicine and pathology, theoretical nuclear physics, evolutionary programming and neural networks, pharmacology, optics, and materials science. Chapter 14 of Stanford and Vardeman 1994 summarizes four diverse Bayesian applications in neutron scattering, nuclear magnetic resonance, global climate change, and rocket photometry. Also, a November 1999 issue of *Science* magazine contains an excellent nonmathematical introductory article (Malakoff 1999) on Bayesian methods in science.

In 1990, there were approximately 3 Bayesian papers per 10,000 scientific publications identified in this database, while in 1999 the rate increased to 12 per 10,000 papers. It may further be shown that this increase in the rate of Bayesian applications exceeded the rate of increase in the overall use of statistical methods in science that likewise occurred during this period. This rapid growth is at least partially due to the recent development of simple Markov Chain Monte Carlo (MCMC) sampling techniques that permit the Bayesian approach to be easily implemented.

Bayesian methods are named for the philosophical approach embodied in the eighteenth-century work of Rev. Thomas Bayes (1702-1761), a Presbyterian minister and mathematician. Bayes' original manuscript on inverse probability (including Bayes' theorem) for the binomial distribution was posthumously published in 1763 (Bayes 1958).

In the frequentist setting, a parameter λ of interest to be estimated is an unknown constant in a specified probability distribution of some corresponding and appropriate measurement *X* (the data). For example, the parameter λ might be the true, but unknown, decay rate (the activity) in an assumed Poisson distribution of an observed number of counts *X* (the measured activity or measurement) in a radioactive sample. In this setting probability represents the long-run frequency with which events of interest involving *X* occur. The only random variability present is in the data *X*. When quantifying uncertainty in an estimate of λ, a frequentist asks questions such as, "Under similar conditions, what other data sets might have been generated? From data set to data set, how much variability would be seen in the estimate of λ? For any one data set, how far might the estimate of λ be from the true value? What is the likelihood of x given that a specified null (versus an alternative) hypothesis is true?"

The Bayesian perspective is quite different. In the Bayesian setting, probability is used to quantify degree of belief. Therefore, λ is assigned a probability distribution. This does not mean that λ varies randomly, only that it is unknown with its probability distribution modeling subjective degree of belief concerning the true value of λ. On the other hand, the data x are treated as observed and therefore fixed. The analyst's prior belief about λ is updated once the data x are obtained, resulting in a posterior belief about λ. A Bayesian asks questions such as, "Given the data, how has my belief about λ changed? What is the current probability distribution of λ given x? What is the probability that λ is contained in a given interval? Given x, what are the posterior odds in favor of a specified null (versus an alternative) hypothesis?

Dr. Corwin L. Atwood, Idaho National Engineering Laboratory, has developed a summary comparison of the major differences between the frequentist and Bayesian approaches. This comparison is shown here in Table 2.1. Each of these notions is discussed in greater detail below.

Table 2.1 Comparison of frequentist and Bayesian approaches to statistical inference.

Topic	Frequentist	Bayesian
Interpretation of probability	Long-run frequency after many hypothetical repetitions	Quantification of degree of belief
Unknown parameter λ	Constant, fixed but unknown	Assigned a probability distribution, measuring our current state of belief
Data *X*	Random (before being observed)	Fixed (after being observed)
Typical estimators for λ	Maximum likelihood estimator (MLE), confidence interval	Bayes posterior mean, credible interval
Interpretation of a 90% interval on λ	If many data sets are generated, 90% of the resulting confidence intervals will contain λ. However, we do not know if our particular interval is one of the lucky ones or not.	We believe the probability is 90% that λ is in our particular interval (in the sense that we would give 9 to 1 odds in a wager).

Both Bayesian and classical methods of inference consider point and interval estimates of unknown parameters based on data. Both methods attempt to precisely identify *a sampling (or measurement) distribution* (or, more precisely, determine the numerical value of one or more parameters of an assumed sampling distribution) based on one or more measurements (sample or measured values) that provide only partial and uncertain information about the distribution. However, in the Bayesian approach, the unknown parameters in the sampling distribution are considered to be values of random variables having a so-called *prior distribution* that expresses the analyst's state-of-knowledge (or subjective probability) about the parameter values prior to considering the current measurements. This formal incorporation of supplementary knowledge and information beyond that contained in the current measurement data is the distinguishing feature of Bayesian inference.

Bayesian inference is defined as those statistical methods in which, prior to the consideration and use of the current sample data, the uncertainty about the values of the parameters of interest in an assumed sampling distribution is appropriately expressed by means of a joint prior distribution.

Let me explain. A sampling distribution is first postulated that relates the unknown parameters to the measurements, and a prior distribution is postulated for the unknown parameters. Before the measurements are used, we are uncertain about the values of the unknown parameters, and our uncertainty is subjectively expressed through the prior distribution. The new information contained in the measurements tells us something about the values of the parameters through the sampling distribution. Because both the prior and the sample measurements contain information about the values of the unknown parameters, we seek to combine them. *Bayes' theorem* provides this combination in the form of a *posterior distribution*, which we will describe in greater detail in Sec. 2.1.1. Even after the prior and sample information has been combined in the form of a posterior distribution, we are still uncertain about the values of the parameters; however, our uncertainty is generally reduced — as it logically should be. The posterior distribution expresses the new updated state-of-knowledge about the parameter values after the sample measurements have been obtained and combined with the prior information. The posterior distribution is then used to make any inferences we desire, such as point and interval estimates of the unknown parameters, calculation of detection limits, etc.

The widespread use of Bayesian methods indicates that scientists and engineers often do indeed have access to, and desire to formally use, relevant prior information. The information is commonly in the form of pertinent objective data such as past measurement results on the same or a similar population of interest, data from related experiments, or less objective (more subjective) data such as scientific conjecture and personal experience.

Bayesian prior distributions must, however, be based on credible analytical analysis. The method cannot be used to put forth a self-serving prior distribution. The prior distribution represents the interpretation of all available evidence, but this interpretation also provides a challenge to communicate the reasonableness of the evidence. In this way, Bayesian methods can add value to a decision analysis process, thereby avoiding misunderstandings.

Bayesian and classical methods differ in one important philosophical respect, which we alluded to above. Classical methods are rooted in the well-known relative frequency notion of probability where the probability of an event is defined as the limiting relative frequency of occurrence of the event in a series of repeated trials. In contrast to this notion of probability, the cornerstone of the foundation of Bayesian methods is the notion of *subjective probability*. Bayesian methods consider probability to be a subjective assessment of the state-of-knowledge (also called degree-of-belief) about propositions of interest, given all available evidence. When the available evidence is mostly empirical, both views often arrive at the same numerical value for the probability in question. For example, a Bayesian would say his subjective probability is one-third that a fair die comes up 1 or 2 when tossed because of what is known about the physics of tossing a die, the uniform mass distribution of the die, and past experience in rolling a die. However, subjective probabilities are appropriate for events that cannot be repeated under similar and controlled conditions, such as the probability of the event that we will have a female president by 2020 or the probability that there is life on Mars. It has been shown that subjective probability has a sound theoretical foundation and obeys the same axioms of probability as relative frequency.

Bayesian and classical methods also differ in their logical approach. Classical statistics uses inductive methods of reasoning; for example, inductive reasoning is used to construct confidence intervals. On the other hand, Bayesian methods use a formal deductive method of reasoning based on Bayes' theorem, which we now introduce in a form pertinent to our needs here.

2.1.1 Prior and Posterior Distributions

Suppose the discrete random variable X has a probability distribution (more formally, a probability mass function) $f_X(\mathrm{x} \mid \lambda)$ which depends on an unknown parameter λ of interest to be estimated. For example, suppose that some measurement X of interest, such as the number of events (e.g., radioactive particle decays) that are observed to occur in a given counting period, has a *Poisson distribution* with parameter λT. In this case, we have

$$f_X(\mathrm{x} \mid \lambda) = (\lambda \mathrm{T})^{\mathrm{x}} \mathrm{Exp}(-\lambda \mathrm{T})/\mathrm{x}!, \mathrm{x} = 0,1,\ldots;\lambda > 0, \tag{2.1}$$

where T denotes the known counting period (or count time) and λ is the unknown rate (per unit time) at which the events occur. Here a measurement consists of observing x events in a specified count time T, and $f_X(x \mid \lambda)$ is the sampling distribution of X. We note here that the sampling distribution is also known as the *likelihood function (or likelihood)* when considered as a function of λ for any observed value x. The likelihood, which is commonly denoted by $\mathcal{L}(\lambda \mid x)$, expresses the information contained in the measured value x regarding λ.

It is also possible to combine data from several independent Poisson processes, each having the same rate λ. In particular, suppose the jth Poisson process is observed for time T_j, yielding an observed measurement x_j. The total number of event occurrences is thus Σx_j in combined (total) counting period ΣT_j. For simplicity, when such data are combined from several processes, we will continue to denote the total count Σx_j by x, and the combined counting period ΣT_j by T.

Further suppose that $\pi(\lambda \mid \theta)$ denotes an appropriate prior distribution for λ that probabilistically expresses our prior belief about λ. Here θ denotes the prior distribution parameters (or *hyperparameters*) in order to distinguish them from the main parameter of interest λ. In other words, while parameters are unknown quantities of interest in sampling distributions, hyperparameters are the parameters of interest in prior distributions. Note that an underlined quantity will be used throughout to denote a vector. For example, in the normal distribution the vector of parameters would be $\theta = (\mu, \sigma^2)$.

For the Poisson sampling distribution in (2.1), $\pi(\lambda \mid \theta)$ must be some suitably chosen probability density function having positive support $(0, \infty)$. For example, prior information is often expressed by means of a so-called *gamma prior distribution*

$$\pi(\lambda \mid \alpha,\beta) = \beta^{\alpha}\lambda^{\alpha-1}\mathrm{Exp}(-\beta\lambda)/\Gamma(\alpha), \lambda,\alpha,\beta > 0, \tag{2.2}$$

where the hyperparameters α and β are the shape and scale parameters of the gamma distribution. In this case, we see that $\theta = (\alpha, \beta)$. Also note that, in order to use (2.2), values for α and β must be chosen in accordance with our prior beliefs about λ. The mean and variance of the gamma distribution in (2.2) are $E[\lambda \mid \alpha, \beta] = \alpha / \beta$ and $V[\lambda \mid \alpha, \beta] = \alpha / \beta^2$, respectively, and values of α and β are often chosen that give desired values of these two moments. Fig. 2.1 illustrates several different gamma distributions plotted using (2.2) for five different pairs of values of α and β. When $\alpha = 1$, note that (2.2) reduces to an *exponential distribution* with decay rate (time constant) β. For the special case in which $\alpha = n / 2$ and $\beta = 1 / 2$, the gamma distribution also reduces to the well-known *Chi-squared (χ^2) distribution* with n degrees of freedom.

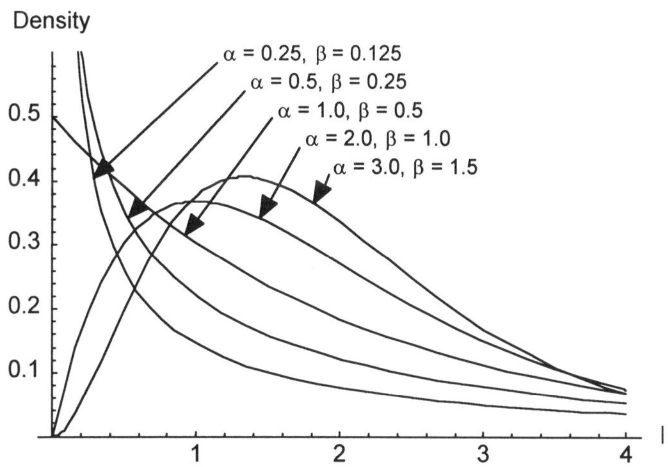

Fig. 2.1 Selected Gamma Distributions With Mean = 2.0.

Bayes' theorem may now be used to calculate the desired *posterior distribution* of λ given the data x; namely,

$$\pi(\lambda \mid x,\theta) = \frac{f_X(x \mid \lambda)\pi(\lambda \mid \theta)}{f_X(x \mid \theta)}, \tag{2.3}$$

where the denominator in (2.3) is $f_X(x \mid \theta) = \int f_X(x \mid \lambda)\pi(\lambda \mid \theta)\, d\lambda$, the so-called *marginal (or unconditional) distribution* of X (that is, unconditional on knowledge of the true, but unknown, value of λ). Note that the denominator is actually the average sampling distribution, averaged over the prior distribution. The denominator normalizes the product of the likelihood and the prior (the numerator) so that the posterior distribution [the LHS of (2.3)] will have an area of one, thus representing a valid probability density function. In other words, Bayes' theorem tells us that the posterior distribution is proportional to the product of the likelihood and the prior, the proportionality constant being the reciprocal of the marginal distribution evaluated at the observed measurement x.

For our Poisson example, it can be shown that the posterior distribution becomes

$$\pi(\lambda \mid x,\alpha,\beta) = (T+\beta)^{x+\alpha}\lambda^{x+\alpha-1}\mathrm{Exp}[-(T+\beta)\lambda]/\Gamma(x+\alpha), \tag{2.4}$$

which we recognize as yet another gamma distribution except that the shape and scale parameters are now $x + \alpha$ and $T + \beta$, respectively. The fact that the prior and posterior are both gamma distributions (but with different parameters) indicates that the gamma distribution is a so-called *conjugate* prior for a Poisson model. Conjugate families of priors are those distributions for which the prior and posterior distributions are members of the same family, such as the gamma family of priors for the Poisson parameter λ that we have considered here. In Sec. 2.2.2 we describe how to use a posterior distribution, such as the gamma posterior in (2.4), to obtain desired inferences and estimates, such as point and interval estimates of λ.

From (2.4), we have a convenient interpretation for the prior shape and scale hyperparameters. We see that α may be interpreted as a prior "measurement" of the event of interest in a prior "count time" β.

Finally, for our Poisson example here, the marginal distribution $f_X(x \mid \theta)$ of X in the denominator of (2.3) is the so-called *gamma-Poisson distribution* given by

$$f_X(x \mid \alpha,\beta) = \frac{\beta^{\alpha}T^{x}\Gamma(x+\alpha)}{\Gamma(\alpha)(T+\beta)^{x+\alpha}\Gamma(x+1)}, \quad x = 0,1,2,.... \tag{2.5}$$

I leave it as an exercise to show that, when we insert (2.1), (2.2) and (2.5) in (2.3), we do indeed obtain the gamma posterior distribution in (2.4).

2.1.2 Dominant Likelihood

We see from Bayes' theorem in (2.3) that the posterior distribution of λ is proportional to the product of the prior and the likelihood. In general, if the prior is relatively flat where the likelihood is appreciable, then the likelihood dominates the prior in the sense that the posterior largely reflects only the information about the parameter(s) of interest contained in the measurements (as expressed by the likelihood). This is known as a dominant likelihood situation. The resulting posterior distribution is then proportional to the likelihood and, in particular, the value of λ that maximizes the likelihood (the so-called *maximum likelihood estimator*) will be the same as the value that maximizes the posterior, the so-called Bayesian *maximum a posteriori* (or MAP) estimator. Otherwise, if the prior is sharply peaked (i.e., a strongly informative prior), then the prior dominates the likelihood, and the posterior largely reflects only the information contained in the prior.

In most cases, the likelihood will increasingly tend to dominate the prior as the quantity of sample data increases. For large sample sizes, the posterior will thus be approximately equal to the normalized likelihood (normalized to have unit area), and the difference between Bayesian and classical inferences will be insignificant. In many practical problems, a moderate sample size will yield a dominant likelihood if the sample results are consistent with the prior distribution. If the sample data are inconsistent with the prior assumptions, then the likelihood will not dominate the prior, and the posterior will reflect a weighted combination of both.

The notion of dominant likelihood is aptly illustrated by a simple coin-tossing experiment. Suppose a coin is flipped n times and heads occurs x times; however, we do not know whether the coin is fair or not. Our problem is to infer the probability p of a head. Fig. 2.2(a) shows three different possible prior states-of-knowledge about p: a uniform (or noninformative) prior (Prior A), a prior that assumes the coin is more likely to be either double-headed or double-tailed than fair (Prior B), and a prior that assumes the coin is probably fair (Prior C).

The likelihood for our coin-tossing experiment is given by the binomial distribution. Fig. 2.2(b)-(d) shows how the posterior for each of the three priors in Fig. 2.2(a) changes as we collect more and more data. Note as we increase the size of n, we become more confident in our inferred value for p (i.e., the variance of each posterior decreases), and the influence of the prior distribution decreases (i.e., the likelihood

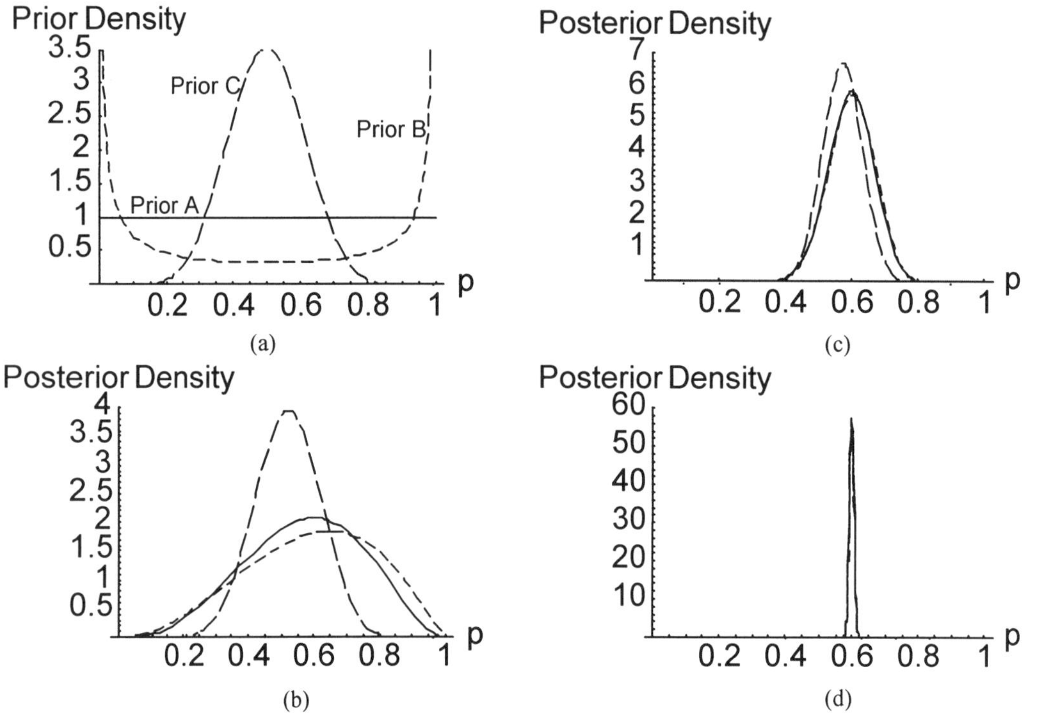

Fig 2.2 Prior and Posterior Distributions for the Coin-Tossing Experiment: (a) Three Different Prior Distributions; (b) Posterior Distributions Given 3 Heads in 5 Tosses; (c) Posterior Distributions Given 30 Heads in 50 Tosses; (d) Posterior Distributions Given 3000 Heads in 5000 Tosses.

increasingly dominates the prior and the posteriors converge to the same answer when sufficient data are available). In other words, no matter what our prior state-of-knowledge, the data force us to the same conclusion. In general, if measurement data are inaccurate, few in number, or inappropriate for the parameter of interest, then the posterior information will crucially depend on the information expressed in the prior.

2.1.3 Advantages and Disadvantages

The primary advantages of Bayesian methods of inference are:

- Bayesian methods provide a logical and unified approach to estimation in that it can measure uncertainty about parameters using probabilities. To the extent that the information embodied in the prior distribution accurately reflects the uncertainty about the parameters, Bayesian methods produce better estimates and decisions than classical estimates.
- Bayesian reasoning is straightforward using deductive reasoning as opposed to indirect inductive reasoning used in classical inference. Furthermore, it has a certain intuitiveness that makes it easy to follow, explain, and appreciate.
- Bayesian hierarchical modeling (Sec. 2.2.3) can be easily used to model and solve complicated problems that otherwise would be extremely difficult, or even impossible, using classical statistical methods.
- Hierarchical Bayesian computations, although tedious, are straightforward and modern computer software exists, such as **BUGS** (Gilks et al. 1994), for performing these calculations.
- Bayesian methods provide a formal method of explicitly introducing prior information and knowledge into an analysis. This is particularly useful when sample data are scarce, such as occurs when an analyst is interested in "needle in a hay stack" events. For example, an analyst may be interested in detecting contaminated workers in a population in which the number of contaminated workers is believed to be small relative to the number of uncontaminated workers.
- While a confidence interval <u>cannot</u> be interpreted as a probability statement about a parameter, probability intervals <u>do</u> have this desired interpretation. They are interpreted as a subjective probability that the parameter is contained in the interval.
- Bayesian intervals using informative priors are often shorter than classical confidence intervals (as we will observe in Sec. 2.3).
- Bayesian point estimates are non-zero even when no Poisson events or binomial successes have been observed.

- Bayesian inference provides a natural and convenient method for use in updating our state-of-knowledge about a parameter as future additional sample data become available; Bayes' theorem is the means used to accomplish this update.
- Bayesian *predictive distributions* provide a simple and convenient way to probabilistically characterize future sample data that have not yet been observed and do not depend on unknown parameters (Gelman et al. 1995).

The primary disadvantages of Bayesian methods are:

- A suitable prior distribution must be identified and justified, which is sometimes a difficult task in practice.
- Bayesian inferences may be sensitive to the choice of a prior distribution.
- A practitioner may find it difficult to convince the technical community at large to adopt his or her prior distribution.
- In this same regard, a practitioner using Bayesian methods is open to the criticism that a prior distribution has been selected that reflects a point of view (or bias) that may be either inappropriate (at best) or self-serving (at worst).

2.2 CLASSICAL, BAYES, HIERARCHICAL BAYES, AND EMPIRICAL BAYES ESTIMATION

In many health physics applications we are interested in obtaining either point or interval estimates (often both) of the unknown parameter(s) of interest in an assumed sampling (measurement) distribution, such as the parameter λ in the Poisson distribution in (2.1). There are four major classes of statistical estimation methods in common use today that can be used to produce the desired estimates. I will discuss each of these methods, emphasizing their major distinguishing features. The four classes are: classical, Bayesian, hierarchical Bayes, and empirical Bayes.

Before discussing these methods, let me introduce a practical Poisson example that I will subsequently use to illustrate each method. A more extensive version of this example was first considered by Martz et al. 1999. When the fission process or reactivity of a nuclear power reactor is above a certain threshold level the reactor is said to be in a *critical state* or, simply, *critical*. One important safety system in a nuclear power plant is the reactor protection system. This system is designed to rapidly change the reactor from a critical to a non-critical status following some transient event, such as loss of offsite power. In response to such an event, control rods, which are made of an appropriate neutron absorbing material, are inserted into the reactor core, thus shutting down the reactivity. This is known as a *reactor trip* or *scram* and can be either automatically or manually accomplished. Reactor scrams can result from initiating events that range from relatively minor incidents to events that are precursors of accidents. The rate at which unplanned scrams occur is thus an important indicator of overall plant performance and reliability.

Table 2.2 contains the unplanned scram data for 1984 for 66 U.S. commercial nuclear power plants each having nonzero critical hours for each year of this period. The data were obtained from NRC report NUREG-1272 1987, and consist of the annual number of unplanned scrams x_i in total critical hours T_i for plant i = 1 ,..., 66. We wish to use this data to estimate the scram rate at each plant. Let m denote the total number of plants considered; thus, m = 66.

We expect the probability of a reactor scram in a sufficiently small time interval to be approximately proportional to the length of the interval and the probabilities of scrams in non-overlapping time intervals to be independent. Thus, the Poisson distribution is an appropriate sampling distribution for describing the sampling variability in the individual plant-specific data in Table 2.2. That is, conditional on the true unknown scram rate λ_i, we assume that X_i follows a Poisson distribution with parameter $\lambda_i T_i$. We further assume that all the X_i values are conditionally statistically independent given their respective true (but unknown) scram rates.

If each plant has the same underlying scram rate, then we should simply pool (combine) the data across plants and use this combined data to estimate the common scram rate. The usual Pearson chi-square statistic can be computed and used to test the equality of λ_i, i = 1, ..., 66. Under the null hypothesis of identical scram rates across plants, this statistic has a chi-square distribution with 65 degrees of freedom. We find that $\chi^2 = 378$, with an associated p-value of nearly 0. Thus, we conclude that the scram rates differ between plants. The main reason for this is that each plant appears to have a unique "identity" as a consequence of different utility operators, different local operating and maintenance procedures, etc.

If the λ_i's are not constant, one approach is to use the data from each plant separately to estimate the scram rate for that plant. Classical estimation methods are the appropriate means to do this (Sec. 2.2.1). However, this approach completely ignores any common aspects of the scram rate that may be jointly shared between plants by virtue of the fact that all U.S. commercial nuclear power plants are built, licensed, and maintained

Table 2.2. U.S. commercial nuclear power plant unplanned scram data for 1984.

Plant	x_i	T_i	Plant	x_i	T_i	Plant	x_i	T_i
Arkansas 1	3	6250.0	Hatch 1	7	5645.2	Prairie Island 1	4	8333.3
Arkansas 2	12	7643.3	Hatch 2	7	3111.1	Prairie Island 2	0	7844.0
Beaver Valley 1	4	6451.6	Indian Point 2	4	4705.9	Quad Cities 1	3	4761.9
Big Rock Point	2	6896.6	Indian Point 3	7	6930.7	Quad Cities 2	2	6896.6
Brunswick 2	3	2654.9	Kewaunee	4	7547.2	Robinson 2	0	616.1
Callaway	12	1503.8	LaSalle 1	9	6293.7	Salem 1	10	2673.8
Calvert Cliffs 1	5	7575.8	LaSalle 2	11	5472.6	Salem 2	10	3389.8
Cook 1	3	8108.1	Maine Yankee	7	6666.7	San Onofre 2	5	5263.2
Cook 2	7	5303.0	McGuire 1	4	6060.6	San Onofre 3	7	5072.5
Cooper Station	3	6000.0	McGuire 2	16	6986.9	St. Lucie 1	6	5555.6
Crystal River 3	2	8333.3	Millstone 1	0	6990.2	St. Lucie 2	9	7377.0
Davis-Besse	4	5555.6	Millstone 2	3	8571.4	Summer	11	5555.6
Diablo Canyon 1	5	1084.6	Monticello	0	810.6	Surry 1	8	5298.0
Dresden 2	3	6521.7	North Anna 1	8	4761.9	Surry 2	14	7446.8
Dresden 3	8	3883.5	North Anna 2	4	6153.8	Susquehanna 1	7	6542.1
Duane Arnold	6	6593.4	Oconee 1	3	7500.0	Susquehanna 2	7	2147.2
Farley 1	2	6896.6	Oconee 2	0	8784.0	Turkey Point 3	8	7339.4
Farley 2	6	8333.3	Oconee 3	4	6557.4	Turkey Point 4	9	5084.7
Fort Calhoun	1	5263.2	Oyster Creek	2	1694.9	Vermont Yankee	2	7142.9
Ginna	1	6666.7	Palisades	1	1562.5	Washington Nuclear 2	23	4364.3
Grand Gulf	7	2089.6	Point Beach 1	0	6420.1	Zion 1	6	6315.8
Haddam Neck	3	6521.7	Point Beach 2	0	7544.2	Zion 2	7	6306.3

to the same NRC regulatory standards. Such commonality is also due to the fact that all plants essentially share the same state of nuclear technology. Thus, we prefer estimation methods that capture the common aspects of scram rate behavior between plants. We anticipate that such methods will produce estimates that are superior to the estimates produced using only the data for each individual plant separately.

What we seek, then, are estimation methods having a basis somewhere between the assumption that the scram rate is the same for all plants and the assumption that the scram rate for a given plant is completely unrelated to the scram rates at other plants. This raises an important question: Do the λ_i's behave as if they were randomly drawn (by nature) from some common (but unknown) prior population? A positive answer leads us to consider both hierarchical Bayesian (HB) and empirical Bayes (EB) estimation techniques (Secs. 2.2.3 and 2.2.4).

2.2.1 Classical Estimation

Perhaps the most common classical point estimate of λ is the so-called *maximum likelihood estimate* (MLE). The MLE is that value of λ that maximizes the likelihood $\mathcal{L}(\lambda \mid x)$; that is, the value of λ that maximizes the probability of obtaining the observed measurement x. For the Poisson case, the MLE of λ is $\hat{\lambda} = x / T$, or, in the case of pooled data, becomes $\hat{\lambda} = \Sigma x_j / \Sigma T_j$. For example, for the data from Arkansas 1 in Table 2.2, we estimate the scram rate to be $\hat{\lambda}_1 = 3 / 6.250 = 0.48$ per 1,000 critical operating hours. In the Poisson case, the estimated *standard error* of $\hat{\lambda}$ (the estimated standard deviation of the MLE) is $x^{0.5} / T$, which, in the case of Arkansas 1, is $3^{0.5} / 6.25 = 0.28$.

We are probably all familiar with the classical notion of a confidence interval estimate in which we provide an interval and claim that we are 95% (say) confident that the interval contains the unknown parameter. However, we have to be careful how we interpret a confidence interval. The precise (true) interpretation is that, in repeated sampling, at least 95% of the intervals so constructed are expected to contain the parameter. What this does not mean is that the probability is at least 95% that any particular interval actually contains the parameter! The reason for this important, but oft misunderstood, distinction is that in order to be able to make such a probability claim, the parameter must be a random variable. In the case of a confidence interval, the parameter is not a random variable.

The equations for use in calculating a 100(1 - γ)% confidence interval for a Poisson rate λ are given in many elementary statistics textbooks (such as Miller and Freund 1977); namely,

$$\left[\lambda_L = \frac{\chi^2_{2x,\gamma/2}}{2T}, \lambda_U = \frac{\chi^2_{2x+2,1-\gamma/2}}{2T}\right], \tag{2.6}$$

where $\chi^2_{\nu,\pi}$ denotes the pth quantile of the χ^2 distribution with ν degrees of freedom. In using (2.6), if x = 0, then simply set $\lambda_L = 0$. However, we see that an upper confidence limit is defined in this case for it is reasonable that observing no occurrences of an event would provide some information about how large λ might be, but not about how small it might be. Also note that the confidence interval in (2.6) is conservative in the sense that the actual confidence level is no smaller than the nominal level of 100(1 - γ)%, although it could actually be larger.

For Arkansas 1, suppose we desire a 95% confidence interval estimate of λ_1. Thus, γ = 0.05. Because $x_1 = 3$, the χ^2 values are $\chi^2_{6,0.025} = 1.237$ and $\chi^2_{8,0.975} = 17.535$. The lower confidence limit is $\lambda_L = 1.237 / (2 \times 6.250) = 0.10$ and the upper limit is $\lambda_U = 17.535 / (2 \times 6.250) = 1.40$. Thus, our desired 95% confidence interval on the unknown scram rate λ_1 is [0.10, 1.40].

2.2.2 Bayesian Estimation

Once the posterior distribution has been obtained we use it to calculate the desired estimates of λ. The most common Bayesian point estimators are the mean, median, and mode of the posterior distribution.

Gelman et al. 1995 and Carlin and Louis 1996 provide a good introduction to Bayesian methods. The Bayesian analog of a classical confidence interval is known as a *credible interval*, and it is obtained directly from the posterior distribution. A 100(1 - γ)% credible interval (λ_L, λ_U) is a probabilistic statement about λ given the sample data x; that is, $P[\lambda_L \le \lambda \le \lambda_U \mid x, \theta] = 1 - \gamma$, where this probability is calculated with respect to the posterior distribution $\pi(\lambda \mid x, \theta)$. Thus,

$$P[\lambda_L \le \lambda \le \lambda_U \mid x,\theta] = \int_{\lambda_1}^{\lambda_2} \pi(\lambda \mid x,\theta)d\lambda = 1 - \gamma. \tag{2.7}$$

Because in the Bayesian case the parameters are considered to be random variables, in contrast to the classical situation, this allows a direct probability statement [as in (2.7)] to be made about λ.

A particularly useful Bayesian credible interval is the so-called *symmetric 100(1 - γ)% two-sided interval* obtained by considering equal probability $\gamma/2$ in each tail of the posterior distribution. The advantage of this particular interval is that it is quite easy to obtain in practice. However, another important and widely used credible interval is the so-called *highest posterior density (HPD) interval* (Box and Tiao 1973). An HPD interval is defined to be the shortest interval among the class of all credible intervals having the desired coverage probability. For illustration purposes, we will consider only symmetric two-sided intervals here.

Consider the Poisson distribution. A special type of gamma prior distribution is obtained if the shape hyperparameter is taken to be $\alpha = 1/2$ and the scale hyperparameter β is taken to be zero. This gives a prior distribution that is proportional to $\lambda^{-1/2}$. This may be seen by ignoring the terms in (2.2) not involving λ (the normalizing constant) and then taking the limit as β -> 0 (with $\alpha = 1/2$). Although this is interpreted as a density function, it is called *improper* because its integral from 0 to ∞ is infinite (and not 1 as required of a density function). Thus, there is no normalizing constant that will permit this to be a proper density function. This particular gamma prior is known as *Jeffreys' noninformative prior distribution* (Box and Tiao 1973). However, from (2.4), the corresponding posterior distribution in this case *is* a proper gamma distribution with shape parameter x + 1/2 and scale parameter T.

Because the mean of a gamma distribution is simply the ratio of the shape and scale parameters, a corresponding Bayesian point estimate of λ is the posterior mean given by (x + 1/2) / T. Note that, even when x = 0, this point estimate can never be zero (as will be the case for the MLE).

Considering the data from Arkansas 1 in Table 2.2, the Bayesian posterior mean point estimate of the scram rate is $\tilde{\lambda}_1 = 3.5 / 6.250 = 0.56$ per 1,000 critical operating hours. Note that this estimate is slightly larger than the MLE, which is 0.48. Likewise, the posterior standard deviation is easily computed to be $(3.5 / 6.250^2)^{1/2} = 0.30$.

It can be shown that the posterior distribution of the transformed random variable $2T\lambda$ given x is a χ^2 distribution with 2x + 1 degrees of freedom. Thus, the corresponding symmetric 100(1 - γ)% two-sided Bayesian credible interval on λ is given by

$$\left[\lambda_L = \frac{\chi^2_{2x+1,\gamma/2}}{2T}, \lambda_U = \frac{\chi^2_{2x+1,1-\gamma/2}}{2T}\right]. \tag{2.8}$$

Although this Bayesian interval is nearly the same as the confidence interval in (2.6), they differ in their respective χ^2 degrees of freedom. As a result, the intervals are numerically similar although the interpretation is completely different. Unlike the classical confidence interval, for the Bayes interval we can claim that, given the data x, $P[\lambda_L \leq \lambda \leq \lambda_U \mid x, \alpha, \beta] = 1 - \gamma$. Although both intervals are numerically close (especially when x is large), the Bayesian credible interval is also slightly shorter at each end.

Again considering the Arkansas 1 data and suppose that we now desire a symmetric two-sided Bayesian 95% credible interval on λ_1. The χ^2 values are $\chi^2_{7,0.025} = 1.690$ and $\chi^2_{7,0.975} = 16.013$. The lower credible limit is $\lambda_L = 1.690 / (2 \times 6.250) = 0.14$, and the upper credible limit is $\lambda_U = 16.013 / (2 \times 6.250) = 1.28$. Thus, we can claim that $P[0.14 \leq \lambda_1 \leq 1.28 \mid x_1{=}3, \alpha = 1/2, \beta = 0] = 0.95$. Note that this interval is 12% shorter than the 95% confidence interval. The Arkansas 1 gamma posterior distribution, mean point estimate, and 95% symmetric two-sided credible interval for λ_1 are illustrated in Fig. 2.3.

2.2.3 Hierarchical Bayes (HB) Estimation

We express the random variation in the scram rate λ_i over the population of plants in Table 2.2 (that is, the plant-to-plant variation in λ_i) by means of the gamma prior distribution $\pi(\lambda_i \mid \alpha, \beta)$ in (2.2). We further assume that, given α and β, the scram rates λ_i are statistically independent. Recall that Bayesian methods require that a prior be completely specified. In contrast to this, both HB and EB methods account for uncertainty in precisely specifying a prior and the prior is determined (fitted) by simultaneous consideration of all the data in Table 2.2. However, they accomplish this in entirely different ways.

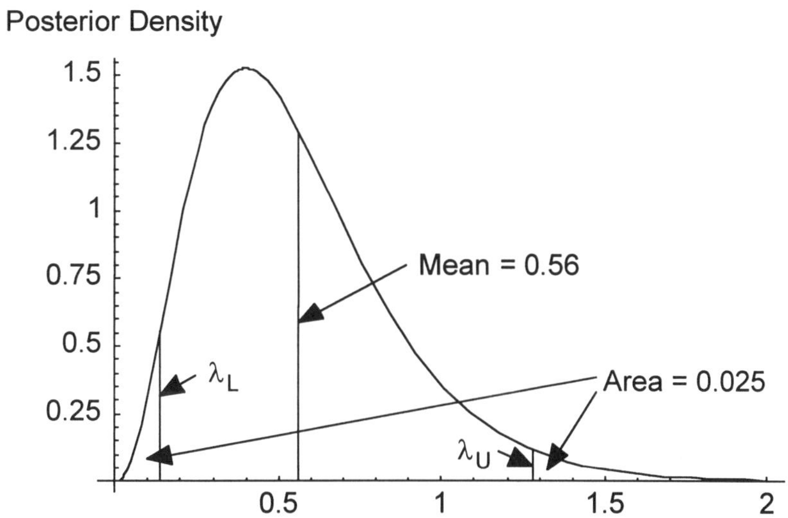

Fig. 2.3 Bayesian Gamma Posterior Scram Rate Density (α = 3.5, β = 6.25) for Arkansas 1.

HB methods account for state-of-knowledge uncertainty about the prior hyperparameters in the form of a *hyperprior distribution* (or second-stage Bayesian prior) on these hyperparameters. For example, in the Poisson case, the hyperparameters α and β in the gamma prior on λ are considered to be random variables with a suitably specified hyperprior distribution. It is usually the case that this hyperprior distribution on α and β is taken to be quite diffuse because we almost never have very precise (or informative) information at this second-stage level of such a model. A Bayesian model in which all the unknown parameters (including the gamma prior distribution hyperparameters α and β) are treated as random variables is known as a *hierarchical Bayes* model. Berger 1985 and Gelman et al. 1995 discuss the basic notions of HB modeling.

Fortunately, we can usually use modern *Markov Chain Monte Carlo (MCMC)* techniques to perform the required calculations in a full hierarchical Bayes model. MCMC is a Monte Carlo integration technique that is implemented using Markov chains. Monte Carlo integration draws samples from the required distribution(s), from which we then form sample averages to approximate expectations. Markov chain Monte Carlo draws these samples by running a cleverly constructed Markov chain for a long period of time. A good introduction to MCMC is provided by Gilks et al. 1996. In our Poisson example, we can obtain the required HB estimates by means of *Gibbs sampling*, a basic MCMC technique that we describe below.

Gibbs sampling is a technique that can be used to generate a random sample from the joint posterior distribution (and hence the marginal posterior distributions) indirectly. The desired HB point and interval estimates can thus be directly computed from the corresponding sample observations obtained via the Gibbs sampler without the need for tedious analytical or numerical calculations.

We briefly describe the method. Suppose that we have a set of p parameters θ_1, θ_2, …, θ_p whose joint distribution $\pi(\theta_1, \theta_2, \ldots, \theta_p)$ is unknown. However, suppose that their full conditional distributions $\pi(\theta_i \mid \theta_j, j \neq i)$ i = 1, 2, …, p, are known in the sense that sample values of θ_i, conditional on values of θ_j, $j \neq i$, may be generated by some method. Under mild conditions, these marginal conditional distributions uniquely determine the required joint distribution $\pi(\theta_1, \theta_2, \ldots, \theta_p)$; hence, all the unconditional marginal distributions $\pi(\theta_i)$, i = 1, 2, …, p, as well.

The Gibbs sampler generates samples from the required joint distribution as follows:

1. Select an arbitrary starting set of values $\theta_1^0, \ldots, \theta_p^0$.
2. Draw θ_1^1 from $\pi(\theta_1 \mid \theta_2^0, \ldots, \theta_p^0)$, then θ_2^1 from $\pi(\theta_2 \mid \theta_1^1, \theta_3^0, \ldots, \theta_p^0)$, and so on up to θ_p^1 from $\pi(\theta_p \mid \theta_1^1, \ldots, \theta_{p-1}^1)$ to complete one iteration of the sampler.
3. After n such iterations, we have obtained the sample $(\theta_1^n, \ldots, \theta_p^n)$. Under mild conditions, this p-tuple converges in distribution to the unknown joint distribution $\pi(\theta_1, \theta_2, \ldots, \theta_p)$ as $n \to \infty$. The set of simulated observations can then be used to estimate $\pi(\theta_1, \theta_2, \ldots, \theta_p)$, its marginal distributions $\pi(\theta_i)$, marginal moments, and so forth.

Fortunately, the above steps in the Gibbs sampler have been conveniently implemented in a widely used software package named "**BUGS**" (Bayesian Inference Using Gibbs Sampling). **BUGS** software (Gilks et al. 1994) and corresponding documentation (including two volumes of worked examples) is available for free download at URL http://www.mrc-bsu.cam.ac.uk/bugs/. Our scram problem is conceptually identical to "Example 2-Pump: Conjugate Gamma-Poisson Hierarchical

Model" in Examples Vol. 1 of the **BUGS** documentation. For our scram example, I will assume that both α and β have rather diffuse independent gamma (0.0625, 0.0625) hyperprior distributions having a mean and standard deviation of 1 and 4, respectively.

Fig. 2.4 contains the **BUGS** computer program used here for our scram example. The file "scram.in" contains the initial values "alpha = 1", "beta = 1", and "lamda[i] = 1, i = 1, ..., 66."

```
model scram;
const
   N=66;                       # number of plants
var
   lamda[N],                   # scram rate for each plant
   x[N],                       # number of scrams for each plant
   t[N],                       # number of critical hours for each plant
   theta[N],                   # lamda[]*t[]
   alpha,beta;                 # gamma prior hyperparameters
data t,x in "scram.dat";
inits in "scram.in";
{
   for (i in 1:N) {
            lamda[i] ~ dgamma(alpha,beta);
            theta[i] <- lamda[i]*t[i];
            x[i] ~ dpois(theta[i]);
   }

   alpha ~ dgamma(0.0625,0.0625);          # mean of 1 and SD of 4
   beta  ~ dgamma(0.0625,0.0625);          # mean of 1 and SD of 4
}
```

Fig. 2.4 BUGS Model Specification for the Scram Example.

After 1,000 burn-in iterations (to remove the effect of the starting values) of the hierarchical model in Fig. 2.4, 10,000 simulated posterior sample values of α, β, and $\lambda = (\lambda_1, \lambda_2, \ldots, \lambda_{66})$ were recorded (p = 68). These 10,000 sample values were then used to calculate the required posterior point and credible interval estimates of α, β, and each λ_i. For example, the HB estimated posterior mean and standard deviation of the Arkansas 1 scram rate are calculated to be 0.59 and 0.28, respectively, per 1,000 critical operating hours. The corresponding symmetric two-sided HB 95% credible interval on λ_1 is [0.17, 1.26]. Note that this interval is slightly shorter than the Bayes interval in Sec. 2.2.2 due to the fact that Arkansas 1 is "borrowing strength" from the scram data at the other plants.

In addition, the marginal posterior mean and standard deviation of α are calculated to be 1.38 and 0.31, respectively, whereas those for β are computed to be 1.21 and 0.32. A symmetric two-sided HB 95% credible interval for α is [0.85, 2.05], while the corresponding interval for β becomes [0.67, 1.90]. The marginal posterior correlation between α and β is also easily calculated to be 0.89.

2.2.4 Empirical Bayes (EB) Estimation

The HB approach considered in Sec. 2.2.3 assumes the existence of a gamma prior distribution for λ_i in the Poisson case. In addition, this prior is further assumed to depend on unknown hyperparameters α and β which in turn follow a second-stage prior distribution. Rather than proceed in this way, the EB approach uses the observed data to directly estimate the unknown hyperparameters α and β as well as the uncertainty in these estimates. These estimates and uncertainties are usually produced using the ML method. When using EB it is important to account for the sampling uncertainties in the hyperparameter estimates when estimating the desired posterior distributions, a fact that we consider below. Maritz and Lwin 1989 and Carlin and Louis 1996 discuss the basic ideas of EB estimation.

The EB estimates of α and β are obtained by maximizing the likelihood function of the data for all m = 66 plants. Because of the assumed conditional independence of the data from each plant, the appropriate likelihood to be maximized is thus a product of the marginal gamma-Poisson distributions in (2.5) (one term in the product for each plant). Thus, we must maximize

$$\mathcal{L}(\alpha,\beta \mid \mathbf{x}) = \prod_{i=1}^{66} \frac{\beta^{\alpha} T_i^{x_i} \Gamma(x_i+\alpha)}{\Gamma(\alpha)(T_i+\beta)^{x_i+\alpha}\Gamma(x_i+1)}, \tag{2.9}$$

as a function of α and β. Here $\mathbf{x} = (x_1, x_2, \ldots, x_{66})$, a vector of counts representing the observed number of scrams from each plant. For the data in Table 2.2, the values of α and β that maximize (2.9) are found to be $\hat{\alpha} = 1.39$ and $\hat{\beta} = 1.21$. From the observed

Fisher information matrix we compute the asymptotic standard deviation of $\hat{\alpha}$ to be 0.32, while that of $\hat{\beta}$ is also 0.32. Moreover, the EB correlation between $\hat{\alpha}$ and $\hat{\beta}$ is 0.89. Note the excellent agreement between these estimates and the corresponding HB estimates in Sec. 2.2.3. Fig. 2.5 shows a plot of the EB-fitted gamma prior distribution.

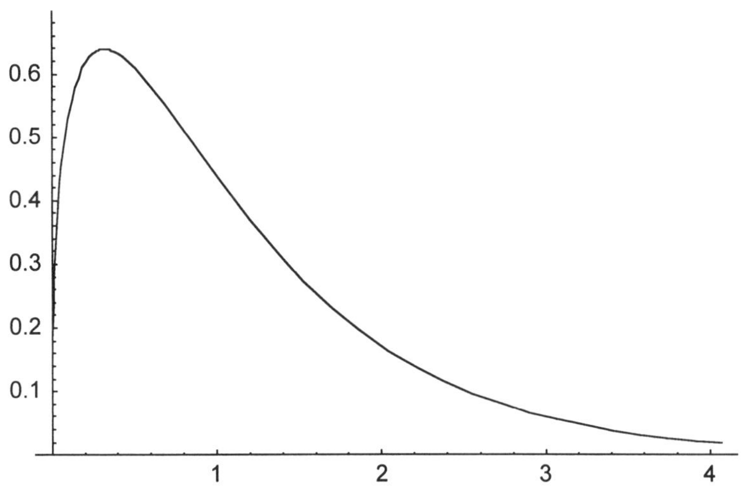

Fig 2.5 Empirical Bayes Fitted Gamma ($\alpha = 1.39$, $\beta = 1.21$) Prior Scram Rate Distribution.

Conover et al. 1997 describe a χ^2 goodness-of-fit (GOF) test for non-identically distributed random variables which is particularly applicable to (2.9). The method they suggest uses the usual χ^2 test statistic that asymptotically follows a χ^2 distribution under the null hypothesis of a gamma-Poisson marginal distribution of x_i. Applying this GOF test to the data in Table 2.2 yields a χ^2 value of 22.34 with 33 degrees of freedom (the associated number of bins is 36) and corresponding p-value of 0.92. Such a large p-value clearly indicates that a gamma-Poisson is an appropriate marginal distribution for the data, thus indirectly supporting the choice of a gamma prior distribution.

The EB estimated marginal posterior distribution of λ_i is a gamma distribution with shape parameter $x_i + \hat{\alpha}$ and scale parameter $T_i + \hat{\beta}$, and this distribution is known to be an approximation to the desired full HB marginal posterior $\pi(\lambda_i \mid \mathbf{x})$. However, this approximation fails to take account of the uncertainty in $\hat{\alpha}$ and $\hat{\beta}$. Although it is well known that the EB estimate of the marginal posterior mean given by $E[\lambda_i \mid x_i, \hat{\alpha}, \hat{\beta}] = (x_i + \hat{\alpha}) / (T_i + \hat{\beta})$ is, to first order, approximately equal to the desired marginal posterior mean $E[\lambda_i \mid \mathbf{x}]$, it is also well known that the marginal posterior variance estimate $V[\lambda_i \mid x_i, \hat{\alpha}, \hat{\beta}] = (x_i + \hat{\alpha}) / (T_i + \hat{\beta})^2$ is too small (because this variance estimate fails to include the statistical uncertainty in the MLEs $\hat{\alpha}$ and $\hat{\beta}$).

We will now broaden this marginal posterior variance to account for the statistical uncertainty in $\hat{\alpha}$ and $\hat{\beta}$ using a method proposed by Kass and Steffey 1989. Applying this method yields the "broadened" marginal posterior variance estimate given by

$$V[\lambda_i \mid \mathbf{x}] \approx \frac{x_i + \hat{\alpha}}{\left(T_i + \hat{\beta}\right)^2} + \frac{1}{\left(T_i + \hat{\beta}\right)^2} V[\hat{\alpha}]$$
$$+ \frac{\left(x_i + \hat{\alpha}\right)^2}{\left(T_i + \hat{\beta}\right)^4} V[\hat{\beta}] - \frac{2\left(x_i + \hat{\alpha}\right)}{\left(T_i + \hat{\beta}\right)^3} Cov[\hat{\alpha}, \hat{\beta}]. \tag{2.10}$$

For an example of how this broadening performs, consider the scram data for Callaway in Table 2.2 ($i = 6$); namely, $x_6 = 12$ and $T_6 = 1.5038$. Note that the MLE is 12 / 1.5038 = 7.98 unplanned scrams per 1,000 critical operating hours and is the largest MLE of the 66 plants. The EB posterior mean estimate of the scram rate is (12 + 1.39) / (1.5038 + 1.21) = 4.93. Because the EB prior mean estimate is 1.39 / 1.21 = 1.15, we note that the EB estimate effectively "shrinks" the MLE down from 7.98 to 4.93, towards the prior mean. For this reason, EB estimators are often referred to as "shrinkage" estimators. The corresponding HB estimator for Callaway turns out to be 5.00, which is likewise a "shrunken" estimate.

I will now illustrate the effectiveness of the "broadening" inherent in (2.10) by again using the Callaway data. The HB marginal posterior variance for the Callaway scram rate is computed using **BUGS** to be 2.16, which I will use as the standard for comparison. From (2.10), the EB "broadened" posterior variance estimate of the Callaway scram rate is calculated to be 2.04. While still an underestimate, it is closer to the HB estimate of 2.16 than the simple "unbroadened" EB posterior variance (12 + 1.39) / $(1.5038 + 1.21)^2$ = 1.82. Thus, although not completely effective, the use of (2.10) is clearly an improvement over no broadening at all.

Again consider the Arkansas 1 data. The EB estimated marginal posterior mean and broadened posterior variance of the Arkansas 1 scram rate are calculated to be 0.59 and 0.0794, respectively, per 1,000 critical operating hours. We can fit a gamma distribution to the true marginal posterior distribution by matching these first two moments to those of a gamma distribution. The corresponding posterior gamma shape and scale parameter values, which we denote by α^* and β^*, are found to be

$$\alpha^* = \frac{Mean^2}{Variance}, \beta^* = \frac{Mean}{Variance}, \tag{2.11}$$

where "Mean" and "Variance" denote the posterior mean and broadened posterior variance given in (2.10), respectively. Thus, in our case, $\alpha^* = 0.59^2 / 0.0794 = 4.38$ and $\beta^* = 0.59 / 0.0794 = 7.43$. By using these values, the corresponding symmetric two-sided EB 95% credible interval on λ_i is calculated to be [0.17, 1.26]. Note that this interval is identical to the corresponding HB interval in Sec. 2.2.3. However, the EB and HB credible intervals are usually not the same.

Table 2.3 contains a summary of the distinguishing features of the Bayes, HB and EB estimation methods.

Table 2.3 A comparative summary of Bayes, hierarchical Bayes, and empirical Bayes methods.

Method	Prior	Posterior
Bayes	$\pi(\lambda\|\theta)$ known, θ known	$\pi(\lambda\|\mathbf{x},\theta) \propto \prod_{i=1}^{m} f_X(x_i\|\lambda_i)\pi(\lambda_i\|\theta)$
Hierarchical Bayes	$\pi(\lambda\|\theta)$ known, θ uncertain	$\pi(\lambda\|\mathbf{x}) \propto \int_\theta \pi(\lambda\|\mathbf{x},\theta)\pi(\theta\|\mathbf{x})d\theta$, where $\pi(\theta\|\mathbf{x}) \propto \pi(\theta)\prod_{i=1}^{m} f_X(x_i\|\theta)$
Empirical Bayes	$\pi(\lambda\|\theta)$ known, θ uncertain	$\pi(\lambda\|\mathbf{x},\hat{\theta}) \propto \prod_{i=1}^{m} f_X(x_i\|\lambda_i)\pi(\lambda_i\|\hat{\theta})$

2.3 RESULTS

In order to compare the results of these four estimation methods, Table 2.4 contains the ordered ML estimates, associated standard errors, and classical 95% confidence interval bounds (Sec. 2.2.1); Bayes posterior mean, standard deviation, and 95% symmetric credible interval limits (Sec. 2.2.2); HB posterior mean, standard deviation, and 95% symmetric credible interval limits (Sec. 2.2.3); and EB posterior mean, standard deviation, and 95% symmetric credible interval limits (Sec. 2.2.4) for the scram rate at each plant.

Both the shrinkage and reduced variability of the HB and EB estimates due to shrinkage are apparent in Table 2.4, particularly for large estimated scram rates. The two Bayes, HB and EB estimates with large uncertainties at the bottom of Table 2.4 correspond to Monticello and Robinson 2 and can be explained. For these two plants no scrams were observed in 1984 in rather small operating times of 810.6 and 616.1 hours, respectively. These small operating times likewise explain the large shrinkage of the HB and EB estimates upwards towards the prior mean of 1.15 relative to the other five plants whose operating times were significantly larger and for which no unplanned scram events were observed.

Table 2.4. Classical maximum likelihood (ML), confidence interval (CI), Bayes, hierarchial Bayes (HB), and empirical Bayes (EB) scram rate estimates

Ordered Plant	ML	Std Error	CI 2.5%	CI 97.5%	Bayes Mean	Bayes StdDev	Bayes 2.5%	Bayes 97.5%
Callaway	7.98	2.30	4.12	13.94	8.31	2.35	4.36	13.51
WashingtonNuclear2	5.27	1.10	3.34	7.91	5.38	1.11	3.43	7.77
DiabloCanyon1	4.61	2.06	1.50	10.76	5.07	2.16	1.76	10.11
Salem1	3.74	1.18	1.79	6.88	3.93	1.21	1.92	6.63
GrandGulf	3.35	1.27	1.35	6.90	3.59	1.31	1.50	6.58
Susquehanna2	3.26	1.23	1.31	6.72	3.49	1.28	1.46	6.40
Salem2	2.95	0.93	1.41	5.43	3.10	0.96	1.52	5.23
McGuire2	2.29	0.57	1.31	3.72	2.36	0.58	1.36	3.63
Hatch2	2.25	0.85	0.90	4.64	2.41	0.88	1.01	4.42
Dresden3	2.06	0.73	0.89	4.06	2.19	0.75	0.97	3.89
LaSalle2	2.01	0.61	1.00	3.60	2.10	0.62	1.07	3.48
Summer	1.98	0.60	0.99	3.54	2.07	0.61	1.05	3.43
Surry2	1.88	0.50	1.03	3.15	1.95	0.51	1.08	3.07
TurkeyPoint4	1.77	0.59	0.81	3.36	1.87	0.61	0.88	3.23
NorthAnna1	1.68	0.59	0.73	3.31	1.79	0.61	0.79	3.17
Arkansas2	1.57	0.45	0.81	2.74	1.64	0.46	0.86	2.66
Surry1	1.51	0.53	0.65	2.98	1.60	0.55	0.71	2.85
LaSalle1	1.43	0.48	0.65	2.71	1.51	0.49	0.71	2.61
SanOnofre3	1.38	0.52	0.55	2.84	1.48	0.54	0.62	2.71
Cook2	1.32	0.50	0.53	2.72	1.41	0.52	0.59	2.59
Hatch1	1.24	0.47	0.50	2.55	1.33	0.49	0.55	2.43
St Lucie2	1.22	0.41	0.56	2.32	1.29	0.42	0.60	2.23
OysterCreek	1.18	0.83	0.14	4.26	1.48	0.93	0.25	3.79
Brunswick2	1.13	0.65	0.23	3.30	1.32	0.70	0.32	3.02
Zion2	1.11	0.42	0.45	2.29	1.19	0.43	0.50	2.18
TurkeyPoint3	1.09	0.39	0.47	2.15	1.16	0.40	0.52	2.06
St.Lucie1	1.08	0.44	0.40	2.35	1.17	0.46	0.45	2.23
Susquehanna1	1.07	0.40	0.43	2.20	1.15	0.42	0.48	2.10
MaineYankee	1.05	0.40	0.42	2.16	1.12	0.41	0.47	2.06
IndianPoint3	1.01	0.38	0.41	2.08	1.08	0.40	0.45	1.98
Zion1	0.95	0.39	0.35	2.07	1.03	0.40	0.40	1.96
SanOnofre2	0.95	0.42	0.31	2.22	1.04	0.45	0.36	2.08
DuaneArnold	0.91	0.37	0.33	1.98	0.99	0.39	0.38	1.88
IndianPoint2	0.85	0.42	0.23	2.18	0.96	0.45	0.29	2.02
Farley2	0.72	0.29	0.26	1.57	0.78	0.31	0.30	1.48
Davis-Besse	0.72	0.36	0.20	1.84	0.81	0.38	0.24	1.71
McGuire1	0.66	0.33	0.18	1.69	0.74	0.35	0.22	1.57
CalvertCliffs1	0.66	0.30	0.21	1.54	0.73	0.31	0.25	1.45
NorthAnna2	0.65	0.33	0.18	1.66	0.73	0.34	0.22	1.55
Palisades	0.64	0.64	0.02	3.57	0.96	0.78	0.07	2.99
QuadCities1	0.63	0.36	0.13	1.84	0.74	0.39	0.18	1.68
BeaverValley1	0.62	0.31	0.17	1.59	0.70	0.33	0.21	1.47
Oconee3	0.61	0.30	0.17	1.56	0.69	0.32	0.21	1.45
Kewaunee	0.53	0.26	0.14	1.36	0.60	0.28	0.18	1.26
CooperStation	0.50	0.29	0.10	1.46	0.58	0.31	0.14	1.33
PrairieIsland1	0.48	0.24	0.13	1.23	0.54	0.25	0.16	1.14
Arkansas1	0.48	0.28	0.10	1.40	0.56	0.30	0.14	1.28
Dresden2	0.46	0.27	0.09	1.34	0.54	0.29	0.13	1.23
HaddamNeck	0.46	0.27	0.09	1.34	0.54	0.29	0.13	1.23
Oconee1	0.40	0.23	0.08	1.17	0.47	0.25	0.11	1.07
Cook1	0.37	0.21	0.08	1.08	0.43	0.23	0.10	0.99
Millstone2	0.35	0.20	0.07	1.02	0.41	0.22	0.10	0.93
BigRockPoint	0.29	0.21	0.04	1.05	0.36	0.23	0.06	0.93
Farley1	0.29	0.21	0.04	1.05	0.36	0.23	0.06	0.93
QuadCities2	0.29	0.21	0.04	1.05	0.36	0.23	0.06	0.93
VermontYankee	0.28	0.20	0.03	1.01	0.35	0.22	0.06	0.90
CrystalRiver3	0.24	0.17	0.03	0.87	0.30	0.19	0.05	0.77
FortCalhoun	0.19	0.19	0.005	1.06	0.28	0.23	0.02	0.89
Ginna	0.15	0.15	0.004	0.84	0.22	0.18	0.02	0.70
Millstone1	0.00	0.00	0	0.53	0.07	0.10	0.0001	0.36
Monticello	0.00	0.00	0	4.55	0.62	0.87	0.001	3.10
Oconee2	0.00	0.00	0	0.42	0.06	0.08	0.0001	0.29
PointBeach1	0.00	0.00	0	0.57	0.08	0.11	0.0001	0.39
PointBeach2	0.00	0.00	0	0.49	0.07	0.09	0.0001	0.33
PrairieIsland2	0.00	0.00	0	0.47	0.06	0.09	0.0001	0.32
Robinson2	0.00	0.00	0	5.99	0.81	1.15	0.001	4.08

Table 2.4. Classical maximum likelihood (ML), confidence interval (CI), Bayes, hierarchical Bayes (HB), and empirical Bayes (EB) scram rate estimates (cont.)

Ordered Plant	ML	HB Mean	HB StdDev	HB 2.5%	HB 97.5%	EB Mean	EB StdDev	EB 2.5%	EB 97.5%
Callaway	7.98	5.00	1.47	2.61	8.29	4.93	1.43	2.54	8.11
WashingtonNuclear2	5.27	4.40	0.91	2.80	6.33	4.37	0.91	2.78	6.33
DiabloCanyon1	4.61	2.83	1.16	1.06	5.52	2.78	1.13	1.02	5.41
Salem1	3.74	2.94	0.88	1.47	4.90	2.93	0.89	1.46	4.91
GrandGulf	3.35	2.56	0.90	1.11	4.58	2.54	0.89	1.10	4.57
Susquehanna2	3.26	2.51	0.88	1.10	4.55	2.50	0.88	1.09	4.49
Salem2	2.95	2.48	0.74	1.25	4.17	2.48	0.74	1.24	4.13
McGuire2	2.29	2.13	0.52	1.25	3.27	2.12	0.51	1.24	3.23
Hatch2	2.25	1.94	0.68	0.85	3.47	1.94	0.68	0.85	3.47
Dresden3	2.06	1.86	0.61	0.85	3.24	1.84	0.61	0.85	3.21
LaSalle2	2.01	1.85	0.53	0.96	3.02	1.85	0.53	0.97	3.03
Summer	1.98	1.85	0.52	0.96	3.01	1.83	0.52	0.95	2.99
Surry2	1.88	1.78	0.45	1.01	2.78	1.78	0.45	1.00	2.77
TurkeyPoint4	1.77	1.66	0.52	0.80	2.82	1.65	0.51	0.80	2.80
NorthAnna1	1.68	1.57	0.51	0.73	2.73	1.57	0.51	0.73	2.73
Arkansas2	1.57	1.50	0.41	0.81	2.41	1.51	0.41	0.81	2.43
Surry1	1.51	1.45	0.47	0.68	2.51	1.44	0.47	0.67	2.50
LaSalle1	1.43	1.39	0.44	0.67	2.37	1.38	0.43	0.67	2.35
SanOnofre3	1.38	1.32	0.46	0.58	2.38	1.34	0.46	0.59	2.38
Cook2	1.32	1.29	0.45	0.56	2.30	1.29	0.45	0.57	2.30
Hatch1	1.24	1.22	0.42	0.54	2.16	1.22	0.42	0.54	2.18
St.Lucie2	1.22	1.21	0.37	0.60	2.05	1.21	0.38	0.59	2.05
OysterCreek	1.18	1.16	0.63	0.27	2.70	1.17	0.64	0.27	2.70
Brunswick2	1.13	1.13	0.54	0.33	2.44	1.14	0.54	0.33	2.42
Zion2	1.11	1.12	0.38	0.50	2.01	1.12	0.39	0.49	1.99
TurkeyPoint3	1.09	1.10	0.36	0.51	1.89	1.10	0.36	0.51	1.91
St.Lucie1	1.08	1.09	0.40	0.45	2.02	1.09	0.40	0.45	2.01
Susquehanna1	1.07	1.08	0.37	0.48	1.92	1.08	0.37	0.48	1.93
MaineYankee	1.05	1.07	0.37	0.48	1.90	1.07	0.37	0.47	1.90
IndianPoint3	1.01	1.03	0.36	0.45	1.85	1.03	0.36	0.46	1.84
Zion1	0.95	0.98	0.36	0.41	1.79	0.98	0.36	0.41	1.81
SanOnofre2	0.95	0.99	0.39	0.37	1.90	0.99	0.39	0.38	1.89
DuaneArnold	0.91	0.94	0.35	0.38	1.74	0.95	0.35	0.39	1.74
IndianPoint2	0.85	0.92	0.40	0.31	1.84	0.91	0.39	0.31	1.83
Farley2	0.72	0.77	0.29	0.32	1.42	0.77	0.29	0.32	1.43
Davis-Besse	0.72	0.79	0.35	0.27	1.60	0.80	0.34	0.27	1.60
McGuire1	0.66	0.74	0.32	0.25	1.48	0.74	0.32	0.25	1.49
CalvertCliffs1	0.66	0.72	0.29	0.27	1.40	0.73	0.29	0.28	1.39
NorthAnna2	0.65	0.73	0.32	0.25	1.47	0.73	0.32	0.25	1.47
Palisades	0.64	0.86	0.57	0.12	2.27	0.86	0.56	0.13	2.25
QuadCities1	0.63	0.73	0.35	0.21	1.55	0.74	0.35	0.22	1.57
BeaverValley1	0.62	0.71	0.30	0.24	1.40	0.70	0.30	0.24	1.41
Oconee3	0.61	0.69	0.30	0.23	1.38	0.69	0.30	0.24	1.39
Kewaunee	0.53	0.62	0.27	0.21	1.23	0.62	0.27	0.21	1.24
CooperStation	0.50	0.61	0.29	0.18	1.29	0.61	0.29	0.18	1.30
PrairieIsland1	0.48	0.56	0.25	0.19	1.14	0.56	0.24	0.19	1.13
Arkansas1	0.48	0.59	0.28	0.17	1.26	0.59	0.28	0.17	1.26
Dresden2	0.46	0.57	0.27	0.17	1.20	0.57	0.27	0.17	1.21
HaddamNeck	0.46	0.57	0.28	0.16	1.21	0.57	0.27	0.17	1.21
Oconee1	0.40	0.51	0.24	0.15	1.08	0.50	0.24	0.15	1.08
Cook1	0.37	0.47	0.23	0.13	1.00	0.47	0.23	0.14	1.01
Millstone2	0.35	0.45	0.22	0.13	0.96	0.45	0.22	0.13	0.96
BigRockPoint	0.29	0.42	0.23	0.10	0.97	0.42	0.23	0.10	0.97
Farley1	0.29	0.42	0.23	0.10	0.97	0.42	0.23	0.10	0.97
QuadCities2	0.29	0.42	0.23	0.09	0.98	0.42	0.23	0.10	0.97
VermontYankee	0.28	0.41	0.22	0.09	0.94	0.41	0.22	0.09	0.94
CrystalRiver3	0.24	0.36	0.19	0.08	0.82	0.36	0.19	0.08	0.82
FortCalhoun	0.19	0.36	0.24	0.06	0.97	0.37	0.24	0.06	0.97
Ginna	0.15	0.30	0.20	0.05	0.78	0.30	0.20	0.05	0.80
Millstone1	0.00	0.17	0.14	0.01	0.54	0.17	0.15	0.01	0.56
Monticello	0.00	0.68	0.59	0.03	2.23	0.69	0.59	0.04	2.23
Oconee2	0.00	0.14	0.12	0.01	0.45	0.14	0.12	0.01	0.46
PointBeach1	0.00	0.18	0.16	0.01	0.57	0.18	0.16	0.01	0.60
PointBeach2	0.00	0.16	0.14	0.01	0.52	0.16	0.14	0.01	0.52
PrairieIsland2	0.00	0.15	0.13	0.01	0.49	0.15	0.13	0.01	0.50
Robinson2	0.00	0.75	0.65	0.03	2.44	0.76	0.65	0.04	2.46

We also note in Table 2.4 that the Bayes, HB, and EB scram rate estimates are non-zero for the seven plants for which no unplanned scrams were reported in 1984. This illustrates an important practical benefit of Bayesian, HB, and EB estimation — the ability to produce a positive, non-zero point estimate even when no events have occurred. Likewise, in contrast to the lower confidence interval bounds, we see that that lower credible interval limits are also non-zero.

We also find in Table 2.4 that the Bayes credible intervals are, on average, 12% shorter than the corresponding confidence intervals, while the HB and EB credible intervals are, on average, both 20% shorter than the corresponding confidence intervals. This is an important result, particularly when we are interested in the precise estimation of a Poisson rate of occurrence.

Finally, the Bayesian, HB, and EB estimation methods that we have introduced here are all potentially useful in other health physics applications. In particular, Chapters 8 and 9 provide further information regarding introductory Bayesian concepts, as well as the use of Bayesian methods in the modeling, analysis, and interpretation of internal dosimetry data.

2.5 ACKNOWLEDGEMENTS

This work was partially supported by the U.S. Nuclear Regulatory Commission, Office of Nuclear Regulatory Research, under Job Code Number W6505 (Dr. Lee R. Abramson, NRC Project Manager) at Los Alamos National Laboratory. I thank Mr. Larry Pratt, Los Alamos National Laboratory, for his assistance in using **BUGS** to obtain the HB estimates in Table 2.4. Finally, I also thank Dr. Michael Hamada, Los Alamos National Laboratory, for proofreading an earlier version of the chapter and for providing several insightful comments.

2.6 REFERENCES

Bayes, T. An essay towards solving a problem in the doctrine of chances. Biometrika. 45:296-315; 1958.

Berger, J. O. Statistical decision theory and Bayesian analysis (second edition). New York: Springer-Verlag; 1985.

Box, G. E. P.; Tiao, G. C. Bayesian inference in statistical analysis. Reading, MA: Addison-Wesley; 1973.

Carlin, B. P.; Louis, T. A. Bayes and empirical Bayes methods for data analysis. London: Chapman & Hall; 1996.

Conover, W. J.; Cox, D. D.; Martz, H. F. A chi-square goodness-of-fit test for non-identically distributed random variables with application to empirical Bayes. Los Alamos National Laboratory Report LA-UR-97-4825, Los Alamos, NM; 1997.

Gelman, A; Carlin, J. B.; Stern, H. S.; Rubin, D. B. Bayesian data analysis. London: Chapman & Hall; 1995.

Gilks, W. R.; Richardson, S.; Spiegelhalter, D. J. Markov chain Monte Carlo in practice. London: Chapman & Hall; 1996.

Gilks, W. R.; Thomas, A.; Spiegelhalter, D. J. A language and program for complex Bayesian modeling. The Statistician. 43:169-178; 1994.

Kass, R. E.; Steffey, D. Approximate Bayesian inference in conditionally independent hierarchical models (parametric empirical Bayes models). Journal of the American Statistical Association. 84:717-726; 1989.

Malakoff, D. Bayes offers a 'new' way to make sense of numbers. Science. 286:1460-1464; 1999.

Maritz, J. S.; Lwin, T. Empirical Bayes methods (second edition). London: Chapman & Hall; 1989.

Martz, H. F.; Parker, R. L.; Rasmuson, D. M. Estimation of trends in the scram rate at nuclear power plants. Technometrics. 41:352-364; 1999.

Miller, I.; Freund, J. E. Probability and Statistics for Engineers (second edition). Englewood Cliffs, NJ: Prentice-Hall; 1977.

Stanford, J. L.; Vardeman, S. B. Statistical methods for physical science. New York: Academic Press; 1994.

U.S. Nuclear Regulatory Commission. Report to the U.S. Nuclear Regulatory Commission on Analysis and Evaluation of Operational Data. Springfield, VA: National Technical Information Service; NUREG-1272, Vol. 2; 1987.

2.7 AUTHOR PROFILE

Harry F. Martz received a Ph.D. in Statistics from the Virginia Polytechnic Institute and State University in 1968. He is presently a member of the Statistical Sciences Group at Los Alamos National Laboratory, and was formerly the group leader for several years. His interests are in Bayesian, hierarchical Bayesian and empirical Bayes applications of statistics, and he has published nearly 100 papers in these areas. He is the principal co-author of the book Bayesian Reliability Analysis and is also a Fellow of the American Statistical Association. His hobbies include collecting southwestern Indian jewelry, treasure hunting, jogging, camping, and traveling.

Chapter 3.0

AN INTRODUCTION TO MONTE CARLO METHODS

James E. Turner

3.1 DEFINITION OF MONTE CARLO

The Monte Carlo method is a technique of numerical analysis that employs random sampling to construct the solution of a mathematical or physical problem. The sampling procedure can be conducted with the help of a series of random numbers r_i, distributed uniformly over the semi-closed interval, $0 \leq r_i < 1$. For many applications, a series can be generated by any one of a variety of algorithms with a digital computer. We briefly examine the generation of random numbers in the next section, before proceeding to a discussion and examples of applications of Monte Carlo procedures to solving problems, with emphasis in health physics. The Monte Carlo technique had its origins at Los Alamos in the 1940s. It is most closely associated with the names, Stanislaw Ulam and John von Neuman, and with the advent of electronic computers. The technique was named Monte Carlo because of the element of chance in choosing random numbers in order to play suitable games for analysis. Selected works on Monte Carlo procedures, random numbers, and random-number generators are given in the reference list (Metropolis and Ulam 1949; Carter and Cashwell 1975; Atkinson 1980; Knuth 1980; Ulam 1983; Kalos and Whitlock 1986; Newman and Barkema 1999). A Monte Carlo primer, written especially for health physicists, is included (Turner et al. 1985).

3.2 RANDOM-NUMBER GENERATORS

A simple random-number generator, used in the early Model-25 Hewlett-Packard hand calculator, is given by

$$r_i = FRAC[(\pi + r_{i-1})^5] \tag{3.1}$$

Here *FRAC*[*x*] denotes the "decimal" part of the argument x that is left when the largest integer less than or equal to x is subtracted from x. To illustrate how (3.1) works, we use an arbitrary seed number, $r_0 = 0.534$, and generate the first five numbers, rounding off each r_i to three significant figures before calculating the next r_{i+1}. We find

$$r_1 = FRAC[(\pi + 0.534)^5] = FRAC[670.868] = 0.868 \tag{3.2}$$

The next numbers in the sequence are

$$\left.\begin{aligned} r_2 &= 0.338 \\ r_3 &= 0.084 \\ r_4 &= 0.179 \\ r_5 &= 0.718\ . \end{aligned}\right\} \tag{3.3}$$

One can see that such a computer-generated, random-number sequence will only go a finite number of steps without repeating itself, commencing when one of the numbers appears for the second time. Since there are only $10^3 = 1000$ different numerical entries possible in the three-place sequence exemplified by (3.3), no such sequence can have a period greater than 1000 entries before repeating itself. Also, since there are only 1000 possible seed numbers, there can be no more than 1000 independent random-number sequences. While hand calculators and electronic computers, especially with double precision, carry many more than three significant figures, their precision is nevertheless finite; and such generators can pose a practical limitation in some Monte Carlo studies. Other types of generators are based on different algorithms, such as use of some aspect of the computer's internal clock reading at various times during program execution. Whereas the numbers thus generated are not periodic in the sense of (3.1), there are still only a finite number of different entries that the sequences in a computer can contain. Because of these and other considerations, when speaking about random numbers and random-number generators in Monte Carlo work, one should perhaps apply the more accurate terminology, "pseudo-random".

Various tests are possible to determine how well a given set of pseudo-random numbers approximates randomness. The distribution of truly random numbers is uniform over the interval $0 \leq r_i < 1$. Their mean, or expected value, is, therefore, $E[r_i] = 1/2$, exactly. Since different entries in a random-number series are completely independent, the expected value of the product of two different entries is $E[r_i r_j] = 1/4$, $i \neq j$. On the other hand, $E[r_i^2] = 1/3$. Also, 10% of the numbers in a random sequence would lie uniformly in each decade of the semi-closed interval [0,1), and so on. Any given sequence of pseudo-random numbers might or might not give a good approximation to some specific criterion for randomness.

3.3 AN EXAMPLE: DETERMINATION OF THE VALUE, π = 3.14159..., BY MONTE CARLO METHODS

3.3.1 Method 1

Figure 3.1 shows a quadrant of a circle inscribed in a unit square. The area within the quadrant, which we shall determine by Monte Carlo means, is $A = \pi/4$. The area A is also the fraction of the area of the square that is inside the circle. If one selects a large number of random points in the square, then the fraction of points that are found to lie within the circular quadrant would have the expected value $\pi/4$. An estimate of π can thus be obtained by counting this fraction from a large sample of such points. Statistically, one would expect that the precise value of π would be approached ever more closely by using more and more points in a sample.

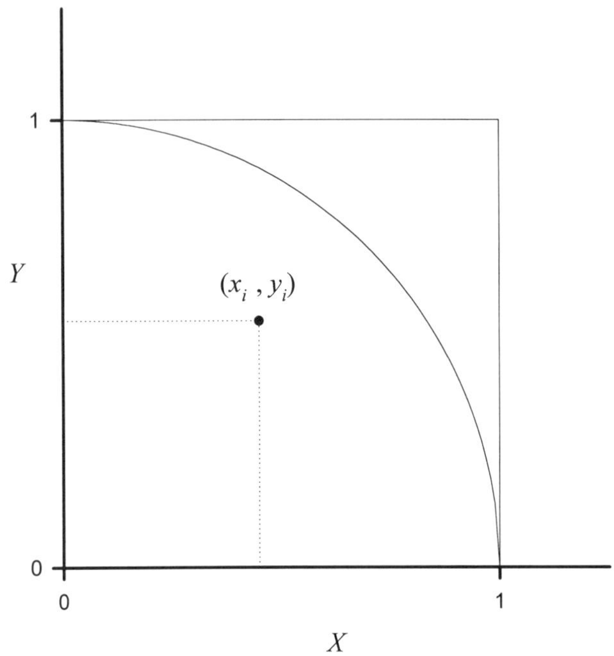

Fig. 3.1 Quadrant of circle inscribed in a unit square has an area $\pi/4$, which is the fraction of random points chosen uniformly in the square that fall inside the circle.

A simple computer program can perform the needed computations. Pairs of random numbers, $0 \leq x_i < 1$ and $0 \leq y_i < 1$, determine the coordinates (x_i, y_i) of points randomly in the square, as indicated in Fig. 3.1. For each point thus selected, a numerical check is made to see whether $x_i^2 + y_i^2 < 1$. When this condition holds, the point is inside the circular quadrant and is tallied; otherwise, the point is noted, but not tallied. After selecting a large number of points, the ratio of the number T of tallied points and the total number N of points used gives the estimated value, $\hat{\pi} = 4T/N$. Numerical results calculated in this way are shown in Table 3.1 for sample sizes ranging from 10 to 10^9 total points. In the sample of 1000 random points, for example, 783 were found to lie within the circular quadrant, implying that $\hat{\pi} = 4(783)/1000 = 3.1320$. This estimate differs from the true value of π by 0.305%.

Table 3.1. Value of π (= 3.14159265...) obtained by Monte Carlo method 1.

Total Number of Points, N	Number of Points, T, within Circle	Estimated Value of π	Percent Error
10	5	2.000000	36.3
10^2	75	3.000000	4.51
10^3	783	3.132000	0.305
10^4	7,927	3.170800	0.930
10^5	78,477	3.139080	0.0800
10^6	785,429	3.141716	0.00393
10^7	7,855,702	3.142281	0.0219
10^8	78,544,278	3.141771	0.00568
10^9	785,385,950	3.141544	0.00155

One sees from the table that using more points does not always give a more accurate computed estimate of π, although there is general improvement as one moves downward in the table. One must remember that the number of tallied points T in column 2 of the table is, itself, a random variable. With a second sample of 1000 points, chosen with a different set of random numbers, one would probably not obtain exactly 783 tallies out of 1000 again. Knowing the answer to this problem in advance, we can say that the expected number of tallies for a sample of 1000 is, in fact, $T = (\pi/4)(1000) = 785.3982$. The probability for the success of any point to be tallied is $p = \pi/4 = 0.7854$. In this method of Monte Carlo computation, the trials for each point to be or not be inside the circle constitute a Bernoulli process (Section 1.2.10.1). The number of successes, or tallies, therefore, is a binomial random variable (Section 1.2.10.2). We can calculate the relative error, or coefficient of variation, for the binomial distribution of T when $N = 1000$. The mean is $\mu = Np$, and the standard deviation is $\sigma = \sqrt{\mu(1 - p)}$. Thus, the coefficient of variation is $\sigma/\mu = 0.0165$, or about 2%, compared with 0.3% shown in the last column of Table 3.1. The particular set of 1000 random points sampled for Table 3.1 had very nearly the expected value of tallies.

Characteristically of Monte Carlo computations, the statistical uncertainty of the results can be assessed, even when we do not know what distribution the random variable satisfies. One can, in principle, always compute the standard deviation for a set of values of T obtained from repeated calculations with a fixed value of N. Conversely, a sample with large N can be subdivided sequentially into smaller samples of equal size, whose distribution in T values would furnish an estimate of the standard deviation associated with the final result.

3.3.2 Method 2

We can, alternatively, compute the area inside the circular quadrant in Fig. 3.1 in another way. The mean-value theorem in calculus states that the area under a curve $y = y(x)$ over an interval of x is equal to the width of the interval times the average value $\bar{y}$ of the curve over that interval. If we find $\bar{y}$ over the unit interval of x in Fig. 3.1, then the area is simply $A = (1)(\bar{y}) = \bar{y}$. Since $A = \pi/4$, we have $\pi = 4\bar{y}$. We proceed with the Monte Carlo solution as follows. For random numbers, $0 \le x_i < 1$ along the X axis, we calculate the corresponding values $y_i = (1 - x_i^2)^{1/2}$, giving the ordinates of the points on the arc at the selected values of the x_i. For a sample of N points x_i, the mean value of y on the arc is

$$\bar{y} = \frac{1}{N} \sum_{i=1}^{N} y_i \tag{3.4}$$

Numerical results for this second method of computation, $\hat{\pi} = 4\bar{y}$, are presented in Table 3.2 for sample sizes from N = 10 to 10^9. The estimate of π in the last column improves with increasing sample size up through 10^5, and then gets steadily worse thereafter. This errant behavior, obtained with the larger samples, occurs because of the limited numerical precision of the computer. It is seen in the second column that the sum of values for the y_i simply "saturates" by the time $N = 10^8$, and does not get any larger as more points are chosen. The problem is one of "roundoff". The computer cannot accurately add two numbers whose difference is greater than its precision; it will give their sum, rounded off, as the larger number alone, as happens in Table 3.2. The sum in column 2 becomes so large that it is insensitive to additional individual contributions y_i, none of which ever exceeds unity in magnitude. [Note that the "saturation" value is exactly equal to $2^{24} = 256^3$, reflecting the binary encoding of the numbers as hexadecimal digits in the computer (Turner et al. 1985).] In recognition of such a limitation, Monte Carlo calculations are often carried out in double precision. (In Table 3.2, the calculations were done in single precision for illustration.) Going to double precision postpones, but does not eliminate, the problem. Roundoff did not occur with method 1, because the variables used there were integers, which are stored differently in the computer than the real (floating-point) numbers used in method 2. In applying Monte Carlo procedures, a good rule is to be alert for lurking computer pitfalls.

Table 3.2. Value of π (= 3.14159265...) obtained by Monte Carlo method 2.

Total Number of Points, N	$\sum_{i=1}^{N} y_i$	$\bar{y}$	Estimated Value of π
10	7.7146697	0.7714670	3.0858679
10^2	80.5046158	0.8050461	3.2201846
10^3	770.8851930	0.7708852	3.0835407
10^4	7,896.6289100	0.7896629	3.1586516
10^5	78,534.3672000	0.7853436	3.1413746
10^6	785,912.8750000	0.7859129	3.1436515
10^7	8,026,959.000000	0.8026959	3.2107835
10^8	16,777,216.0000000	0.1677722	0.6710886
10^9	16,777,216.0000000	0.0167772	0.0671089

3.4 RADIATION TRANSPORT CALCULATIONS

In this section we sketch the elements of some typical Monte Carlo approaches for calculating the transport of radiation through matter. Monte Carlo methods find widespread applications for determinations of absorbed dose, dose equivalent, and shielding. In such computations, one would like to make the computer-generated particles or photons do statistically exactly what they do in nature. A number of Monte Carlo computer codes are available from laboratories and distribution centers for calculating the transport of various kinds of radiation in numerous materials.

3.4.1. Photons

In a typical application, the goal is to determine the spatial distribution of absorbed dose in a tissue phantom exposed to X or gamma rays, having specified distributions of energy and spatial incidence. To carry out the computations, one randomly selects from these given distributions a large number of incident photons. For each, one calculates a complete transport history through the phantom, following the photon and any secondary radiation it produces until it is absorbed or escapes from the phantom. One records all events and their locations in the phantom; and then analyzes the histories of all of the photons, compiling the dosimetric information desired. The first step in this process is to choose a flight distance to the site of the photon's first interaction (if any) in the phantom. This choice is made on the basis of a random number used in conjunction with the linear attenuation coefficient for the given photon energy and material of the phantom. This coefficient is usually an experimentally known quantity, which can be measured under conditions of "good geometry" (Turner 1995, pp. 185-188). It describes the attenuation of a pencil beam of incident, monoenergetic photons as they penetrate matter. The relative number that travel a distance $X > x$ in a material without interacting is found to be described by the familiar exponential function, $e^{-\mu x}$, in which μ is the attenuation coefficient, having the inverse units of x. Since it gives the *relative* number, the exponential thus represents the *probability* that a given incident photon will travel a distance greater than x without interacting:

$$P[X > x] = e^{-\mu x} \tag{3.5}$$

The constant μ can be interpreted as the probability of interaction per unit distance traveled for the photon, and $1/\mu$ is the mean free path for the photon in the given material. In this regard, there is a complete analogy between Eqn. (3.5) and the familiar exponential radioactive decay law for a pure radionuclide. The fraction of the initial number of atoms that have not decayed at time t is $e^{-\lambda t}$, the decay constant λ is the probability of decay per unit time for an atom, and $1/\lambda$ is the mean life.

To gain some insight into how the attenuation coefficient statistically governs the penetration of photons in matter, we look at a specific example. For 300-keV photons in soft tissue, $\mu = 0.118\ \text{cm}^{-1}$. This number, together with Eqn. (3.5), enables us to answer a variety of questions about photons of this energy in tissue. For instance, the probabilities that a given photon will travel a distance of at least 1, 5, or 6 cm without interacting are (unless stated otherwise, we shall use cm as the unit of length throughout)

$$P[X > 1] = e^{-0.118 \times 1} = 0.889 \tag{3.6}$$

$$P[X > 5] = e^{-0.118 \times 5} = 0.554 \tag{3.7}$$

and

$$P[X > 6] = e^{-0.118 \times 6} = 0.493 \tag{3.8}$$

The probability that it will have its first collision between $x = 5$ and $x = 6$ cm is equal to the amount that the penetration probability is diminished between the two distances:

$$P[5 < X < 6] = P[X > 5] - P[X > 6] = 0.554 - 0.493 = 0.061 \tag{3.9}$$

The penetration can be understood as the product of incremental probabilities. Using the probability for traveling each cm, we have for the probability of going at least 5 cm without interacting,

$$P[X > 5] = (P[X > 1])^5 = (0.889)^5 = 0.555 \tag{3.10}$$

which agrees with Eqn. (3.7) to within roundoff in the computations. Thus, one sees that the numerical result (3.7) follows immediately form (3.6). Exponents add in multiplication. This special property of the exponential distribution is referred to as *independent increments*. A photon thus travels without memory of how far it has already gone.

The probability that an incident photon will interact between 5 and 6 cm can also be regarded as the product of the probability $P[X > 5]$ that it reaches 5 cm and the probability $1 - P[X > 1]$ that it then *will not* travel an additional cm without interacting:

$$P[5 < X < 6] = P[X > 5](1 - P[X > 1]) = 0.554(1 - 0.889) = 0.061 \qquad (3.11)$$

in agreement with (3.9). The probability that an un-collided photon at any depth travels an additional cm without interacting is always $P[X > 1]$. This feature of radiation penetration can also be described as follows. As shown by Eqn. (1.5), the conditional probability of an event A, given that another event B has already occurred, written $P[A|B]$, is given by

$$P[A|B] = \frac{P[A,B]}{P[B]} \qquad (3.12)$$

where $P[A,B]$ denotes the joint probability that both A and B occur. Thus, the probability that a photon, having already penetrated 5 cm (event B), will travel an additional cm (event A) is

$$P[X > 1|X > 5] = \frac{P[X > 1, X > 5]}{P[X > 5]} = \frac{P[X > 1]P[X > 5]}{P[X > 5]} = P[X > 1] = 0.889 \qquad (3.13)$$

Because the events are independent, $P[X > 1, X > 5] = P[X > 1]P[X > 5]$, as described by Eqn. (1.5). Regardless of how far a photon has already traveled, or how long it has "been at risk" for having a collision, it always has the same probability of interaction for the next cm of travel.

Still other questions can be answered directly from knowledge of the attenuation coefficient. What is the probability that an incident photon will *not* interact between 5 cm and 6 cm? It follows directly from Eqn. (3.9) that the answer is $1 - 0.061 = 0.939$. Looked at in another way, though, the desired probability is the sum of (1) the probability $1 - P[X > 5]$ that the photon *does* interact between 0 and 5 cm and (2) the probability $P[X > 6]$

that it penetrates at least 6 cm. Thus,

$$1 - P[X > 5] + P[X > 6] = 1 - 0.554 + 0.493 = 0.939 \qquad (3.14)$$

as follows also from Eqn. (3.9).

We next show how one can choose a random flight distance for an incident photon to have its first collision in a target. The probability $f(x)\mathrm{d}x$ that the first collision occurs in a small interval between x and $x + \mathrm{d}x$ is the product of the probability $\mathrm{e}^{-\mu x}$ that the photon reaches the depth x and the probability $\mu \mathrm{d}x$ that it interacts in the next distance $\mathrm{d}x$:

$$f(x)\mathrm{d}x = \mu \mathrm{e}^{-\mu x}\mathrm{d}x \qquad (3.15)$$

The function $f(x)$ is called the probability density function (Section 1.2.5), or PDF, for the first-collision sites X. One sees from Eqn. (3.15) that $f(x)$ has unit area over the range $[0,\infty)$ of the random variable, as required of a PDF. [When $x < 0$, $f(x) = 0$.] The probability that the first collision takes place at a depth less than x is

$$F(x) = \int_0^x f(x')\mathrm{d}x' = \mu \int_0^x \mathrm{e}^{-\mu x'}\mathrm{d}x' = 1 - \mathrm{e}^{-\mu x} \qquad (3.16)$$

This function, which is called the cumulative distribution function (Section 1.2.6), or CDF, is shown in Fig. 3.2 for 300-keV photons in soft tissue ($\mu = 0.118$ cm^{-1}). It has the value $F(x) = 0$ for negative values of x and increases monotonically toward unity as x increases without limit. It follows from (3.16) that the probability for the first collision to occur at a depth between x_1 and x_2, where $x_2 > x_1$, is

$$P[x_1 < X < x_2] = F(x_2) - F(x_1) \qquad (3.17)$$

Equation (3.17) is a generalization of Eqn. (3.9) for finding the probability that the first collision takes place between two arbitrary values of X. As shown in Fig. 3.2, the probability for the collision to be in the interval $x_1 \le x < x_2$ is equal to the fraction of the unit interval along the ordinate that determines the values of x_1 and x_2. Thus, if we select a large number of random points $0 \le r_i < 1$ uniformly along the ordinate and use the cumulative function, $F(x)$ given by Eqn. (3.16), to calculate flight distances x_i, the latter should (statistically) be exponentially distributed, as measured in accordance with Eqn. (3.5). Substituting $F(x) = r_i$ and $x = x_i$ in Eqn. (3.16) and solving for x_i, we obtain

$$x_i = -\frac{1}{\mu}\ln(1 - r_i) \tag{3.18}$$

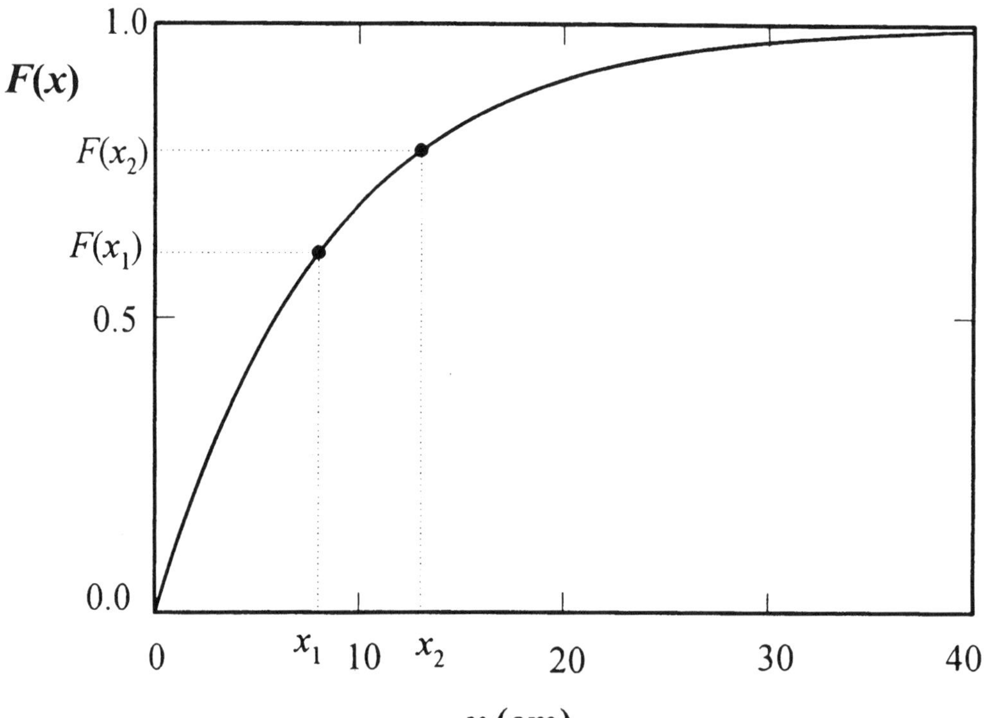

Fig. 3.2 Probability for the flight distance to site of first collision to be in the interval $x_1 \le x < x_2$ is equal to the difference in the cumulative distribution function $F(x_2) - F(x_1)$.

Using this relationship with a series of random numbers will produce a sample of flight distances x_i to the first-collision sites for photons with linear attenuation coefficient μ in accordance with Eqn. (3.5).

The procedure just described shows, in general, how one can use Monte Carlo methods to generate a random sample of events from a given PDF (or from a given discrete probability distribution). As we did with the density function $f(x)$ from Eqn. (3.15), one forms the cumulative distribution. Like $F(x)$ in Eqn. (3.16), the CDF spans the range of values $0 \le F(x) < 1$ over the entire range $-\infty < x < \infty$. Using a series of random numbers, $0 \le r_i < 1$, where $i = 1, 2, ..., N$, one solves each equation,

$$F(x_i) = r_i \tag{3.19}$$

for x_i in order to generate the sample of N values of x_i from the original PDF, as we did with Eqn. (3.18).

For calculating the spatial distribution of absorbed dose in a phantom, the latter can be subdivided mathematically for analysis into a number of sub-volumes. When an incident photon's first-collision site is determined, the program notes the particular sub-volume in which the event occurs. The type of event that takes place there can then also be selected statistically. A 50-keV photon, for instance, that interacts in soft tissue is either photoelectrically absorbed 88.4% of the time or Compton scattered 11.6% of the time (again, experimental data). A random number can be used to determine which of these two alternatives occurs for the interacting photon. For example, if the random number is less that 0.884, then the event can be assigned as photoelectric absorption; otherwise, it is Compton scattering. In either case, a typical Monte Carlo program goes on to compute as much physical detail as desired by means of additional statistical algorithms. Following photoelectric absorption, one can simply tally the photon's energy as energy absorbed locally in the volume element and then proceed to the next incident photon. Or one can program, in addition, the transport of the ejected photoelectron away from the collision site as well as the probabilities for the ejection of Auger electrons and fluorescence photons in response to the inner-shell vacancy created in the struck atom. Analogous events can be added for Compton scattering. The phantom itself can be comprised of diverse materials in various geometrical shapes. The inclusion of very general conditions and complete physical detail is an inherent and valuable feature of Monte Carlo radiation-transport computations. Naturally, when more details and alternatives are considered, longer computational and analysis times are required. In all stages of the computations, experimentally measured values should be used for the input data to the codes insofar as possible. Theoretical approximations are less desirable, but often unavoidable.

After calculating the complete histories of a large number of incident photons, the total energies absorbed in the various volume elements provide a numerical assessment of the spatial distribution of the absorbed dose in the phantom. A number of practical and judgmental considerations enter the design of the computer program to carry out the computations. In addition to the amount of physical detail to include, the analysis of the photon histories must be carefully planned for optimum results. If a small number of sub-volumes comprise the phantom, then the spatial resolution of the calculated absorbed dose in the phantom for a given number of incident photons will be crude. However, the number of events in each sub-volume will be relatively large, and hence the variances of the doses computed there will be relatively small. For better spatial resolution, a larger number of sub-volumes can be used to span the phantom. Then, however, with the same number of photon histories there will be on average fewer events per volume element, leading to greater statistical fluctuations in the raw calculated results. The computer time needed for analysis of a given number of histories is also increased when there are more sub-volumes. As a general rule, Monte Carlo dose calculations for any kind of radiation require practical trade-offs and careful consideration of the physical processes involved, the specific data being sought, and the degree of statistical confidence desired.

3.4.2 Neutrons

Monte Carlo calculations for neutron transport and dose in a phantom have many similarities with those for photons. The flight distance to the first collision for a neutron of given energy in a material is governed statistically by Eqn. (3.5), in which μ is the macroscopic cross section, $1/\mu$ being the mean free path for the neutron in the material. Given the numerical value of μ, Eqn. (3.18) can be used to provide a random flight distance to the first-collision site for a neutron. Neutrons interact with matter through the strong, or nuclear, force with atomic nuclei. When a collision occurs, a random choice of the type of nucleus struck can be made next. In a soft-tissue phantom, the choice would ordinarily be from among the four elements H, C, O, and N, which make up 99+% of the target. These elements contribute specified relative amounts to the macroscopic cross section, μ, and so a random number, $0 \le r_i < 1$, can serve to pick one of them for the collision, consistent with the fraction it contributes to μ. The type of interaction can be selected next: primarily (depending on the neutron energy) elastic scattering, inelastic scattering, absorption, or a reaction, such as (n,p). The probabilities for all events as functions of neutron energy can be conveniently stored as numerical tables with suitable mesh sizes to enable efficient interpolation. Charged and uncharged secondary particles and photons produced by the neutron interactions must also generally be transported in dose and shielding calculations.

In contrast to photons, neutron absorbed-dose computations usually include an assessment of the spatial distribution of dose equivalent in a phantom. To accomplish this, the energy deposited from every microscopic event has to be weighted by an appropriate quality factor. The latter is defined as a function of the linear energy transfer, or LET, for the charged particles involved. For many purposes, the simple use of average quality factors, obtained from published detailed Monte Carlo computations for neutrons of given energies, suffices.

3.4.3 Charged Particles

Energetic charged particles experience primarily the electromagnetic force of interaction with the atomic electrons in the matter they traverse. (The elastic scattering of slow, heavy charged particles by nuclei is usually a minor effect, but can be included in a Monte Carlo program.) In addition, an energetic electron can undergo an abrupt deflection by a nucleus, thereby emitting bremsstrahlung, or continuous X radiation. As with photons and neutrons, the numerical probabilities for all relevant events as functions of energy serve as input data for the Monte Carlo transport program.

Depending on the objectives sought, different approaches can be employed to compute dose by Monte Carlo means. Perhaps the simplest can be described as "ray tracing". An example is the computation of dose from a distributed alpha-particle source, such as ^{226}Ra, deposited in a specific configuration in a bone structure. An alpha particle, emitted from a random point and in a random direction, is assumed to travel in a straight line, slowing down continuously until it stops. Its rate of energy loss along this trajectory is determined by the stopping power ($-dE/dx$) of the medium. The passage through different tissue regions with different stopping powers can be readily followed in the computations. The energy deposited as well as the LET and dose equivalent for a large number of rays can be compiled in different sub-volumes as functions of location. Use of the continuous slowing-down approximation ignores energy-loss fluctuations along the trajectory, which are unimportant in many applications.

Alternatively, one can perform a full, microscopic Monte Carlo calculation, treating each individual collision of a primary charged particle with an atomic electron in a manner analogous to that described above for collisions of photons and neutrons. However, charged-particle collisions take place much more frequently and on a much smaller distance scale than before. For example, we see from Eqn. (3.6) that the mean free path for 300-keV gamma rays in soft tissue is $1/\mu = 1/(0.118\ cm^{-1}) = 8.5$ cm; the mean free path for 100-keV neutrons is 1.1 cm. In contrast, the mean free path for 10-keV electrons in soft tissue is only about 1.3×10^{-6} cm. A charged particle typically loses on the order of 20 eV to perhaps 50 eV in an electronic collision. It *can* lose considerably more (or less), but such events are not frequent. Thus, a 5-MeV alpha particle, for instance, might undergo some 200,000 collisions in coming to rest. Monte Carlo calculations with this amount of detail require a great deal of computer time, and they also often include transport of the secondary and later-generation electrons that are produced.

A primary goal for carrying out full-scale Monte Carlo computations of charged-particle tracks with their detailed structure is to study the basic physical and chemical events that occur and the mechanisms that could be responsible for producing the biological effects of ionizing radiation on living tissue. We cite only two recent examples of such computations here. The production of free radicals and other chemically reactive species and the calculation of their time-dependent yields, measured in liquid water and in aqueous solutions, have been investigated at a number of laboratories. Such computations include not only the transport of the primary radiation and secondary electrons, but also identification of the specific quantum states of the ionized and excited species produced in the water. The calculations follow the subsequent temporal development of these states into chemically active radiolytic products. The latter are programmed to diffuse randomly and react among themselves to produce observable chemical yields. One study, including both Monte Carlo simulation and experimental measurements, was made of the radiolysis of the dipeptide, glycylglycine, irradiated with X rays (Bolch et al. 1998). The computations were performed for the specific conditions under which the yields of a variety of chemical products were measured. In matching the observed yields, the Monte Carlo model provides a detailed accounting in time of all individual species and their competing reaction pathways. While such agreement does not constitute validation of the model, it provides some assurance that it is probably generally correct in major ways. Another set of Monte Carlo calculations have been carried out for targets of DNA structures at various levels of complexity, from single strands to chromosomes (Friedland et al. 1999). The interplay of literal Monte Carlo modeling of microscopic events and the comparison of calculated results with laboratory measurements is a powerful research tool.

3.5 ISOTROPIC CHORD-LENGTH DISTRIBUTION IN A SPHERE

3.5.1 Introduction

The subject of chord-length distributions in cavities affords a basic approach to the technical problem of determining LET spectra for radiation from measurements of energy-proportional pulse-height spectra. Knowledge of LET is deemed relevant to understanding the biological effects of radiation, and it is also required for the assignment of quality factors for dose equivalent. The pioneering work of Rossi and coworkers in the 1950s and later brought this subject and the field of microdosimetry into being (Rossi 1968; Rossi and Zaider 1996).

3.5.2 LET, Chord Lengths, and Microdosimetry

The basic idea for obtaining LET spectra is this. Consider a proportional counter, having a spherical gas cavity (sensitive volume) that is traversed uniformly and isotropically by charged particles. Ideally, the gas pressure and cavity size are made so that the passage of a particle takes place at constant LET, L, along the chord that it traces through the cavity. To a first approximation, the energy ϵ registered from the passage of each particle is equal to the product of the LET for that particle and the length η of the chord traversed:

$$\epsilon = L\eta \tag{3.20}$$

It follows that an observed pulse of a given size ϵ might be produced by a high-LET particle traveling over a short chord or a low-LET particle over a long chord. The distribution of chord lengths under the stated conditions is known (the reason for choosing a spherical cavity). The probability that a chord has a length between η and $\eta + \mathrm{d}\eta$ is

$$f(\eta)\mathrm{d}\eta = \frac{\eta}{2R^2}\mathrm{d}\eta \tag{3.21}$$

where $f(\eta)$ is the PDF, R is the radius of the sphere, and $0 \le \eta < 2R$. This function, which is linear in the chord length, is shown in Fig. 3.3. Statistically by means of Eqn. (3.21), one can associate a probability for chord length η randomly with each pulse of a given size in a measured pulse-height spectrum. Thus one can, in principle, unfold a spectrum of values of $L = \epsilon/\eta$ from the measured pulse-height spectrum of values of ϵ. (Mathematically, the problem is equivalent to solving a convolution integral equation.)

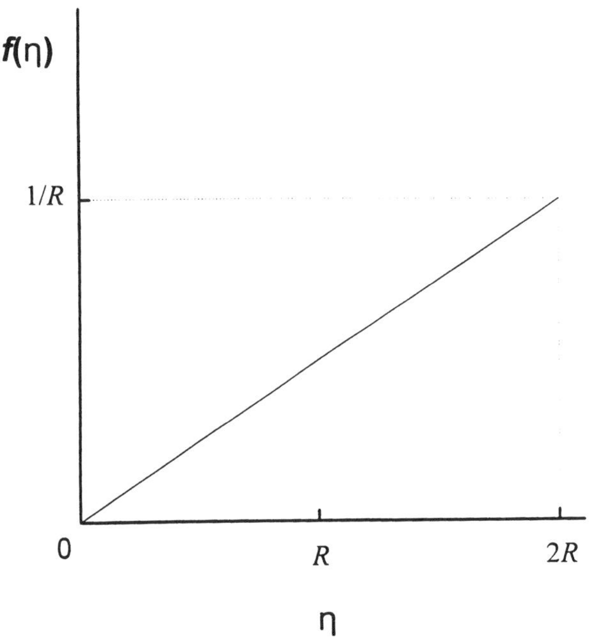

Fig. 3.3 Analytic probability density function for uniform, isotropic chord-lengths η in a sphere of radius R is $f(\eta) = \eta/(2R^2)$.

In practice, however, considerable technical skill is needed to make the system work. In addition to the instrumentation required, Eqn. (3.20) is never a particularly good approximation to reality. Rather than having a fixed value at low gas pressure, ϵ is a highly skewed distribution for a particle of given L and η. At high pressures, L is not constant over the entire chord length; and the cavity also contains "stoppers" and "starters" in addition to "traversers". Nonetheless, spherical, tissue-equivalent Rossi counters have played a valuable role in radiation monitoring. In addition, with proper selection of sphere size and gas pressure, one can employ such chambers to simulate unit-density spheres that have relevant biological sizes, e.g., that of a cell, cell nucleus, or other microscopic structure. The subject of microdosimetry deals with the distribution of physical events on a scale of microns and smaller and the use of such distributions to interpret and understand the biological effects of radiation (Rossi and Zaider 1996).

3.5.3 Chord Lengths in a Sphere

In Section 3.4.1 we showed how Monte Carlo procedures can be used to provide a random sample of events from a given probability density function by means of Eqn. (3.19). One can also use Monte Carlo methods to go in the opposite direction and find an unknown probability density function, at least numerically. Let us suppose, for example, that we did not know what the chord-length distribution in Fig. 3.3 looks like, but would like to determine it. The solution to this problem by Monte Carlo means is described next.

For the task at hand, one can mathematically insert a large number of uniform, isotropic lines at random into a sphere; compute the length of each chord inside the sphere; and compile the distribution of lengths found. Choosing the entry point into the sphere is simplified if, instead of the isotropic distribution in three dimensions, one uses a uniform, parallel beam of lines in two dimensions. Because of spherical symmetry, the same chord-length distribution results. As seen from a side view of such a beam in Fig. 3.4(a), the length of a sample chord at a distance ρ_i from the center of a sphere of radius R is

$$\eta_i = 2\sqrt{R^2 - \rho_i^2} \tag{3.22}$$

In Fig. 3.4(b), which is seen in the direction of the beam, choosing an entry point into the sphere can be accomplished by randomly selecting the coordinates (x_i, y_i) of a point in the circle of radius R, similarly to what was done in Fig. 3.1. The value of ρ_i is found from x_i and y_i by the Pythagorean theorem, as indicated in Fig. 3.4(b); and the chord length is then computed from Eqn. (3.22).

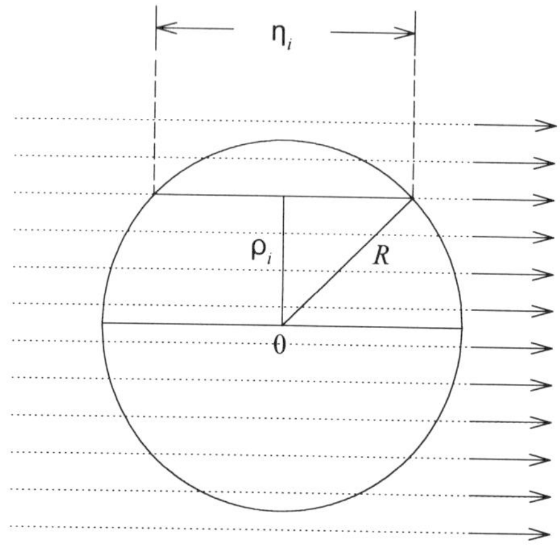

$$(\eta_i/2)^2 = R^2 - \rho_i^2$$

(a) Side View of Beam

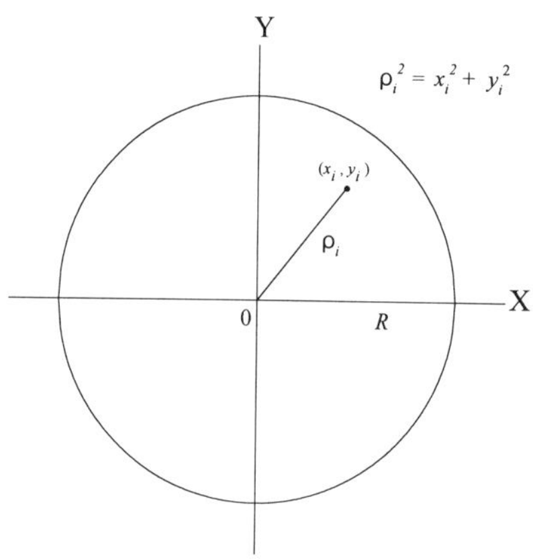

(b) End View (Beam into Page)

Fig. 3.4 **(a)** Broad, uniform beam of parallel tracks traversing sphere of radius R centered at O. **(b)** Random point in circle can be used to direct a random chord through the sphere.

It is convenient to express the chord lengths as dimensionless fractions of the longest chord, whose length is the diameter of the sphere, $2R$. We rewrite Eqn. (3.22), dividing both sides by $2R$:

$$\frac{\eta_i}{2R} = \sqrt{1 - 4\left(\frac{\rho_i}{2R}\right)^2} \tag{3.23}$$

Letting $\hat{\eta}_i = \eta_i/(2R)$ and $\hat{\rho}_i = \rho_i/(2R)$ denote the dimensionless quantities, we write

$$\hat{\eta}_i = \sqrt{1 - 4\hat{\rho}_i^2} \tag{3.24}$$

Because of symmetry, no generality is lost by always choosing the random point (x_i, y_i) in the first quadrant of the circle in Fig. 3.4(b). Then the values of x_i and y_i will vary in magnitude between 0 and 1/2. Therefore, random points of entry can be selected by using the algorithm,

$$x_i = 0.5r_i \qquad \text{and} \qquad y_i = 0.5r_i' \tag{3.25}$$

where $0 \le r_i < 1$ and $0 \le r_i' < 1$ are random numbers. If a selected point falls outside the circle, it is not used.

It is instructive to present a simple calculation with some handling of the statistical data which exemplifies many Monte Carlo studies. For illustration, a random-number generator was used to produce a small sample of 25 points in the first quadrant of the circle, as shown in Fig. 3.4(b). For each point, the resulting chord length was calculated from Eqn. (3.24), thus providing a random sample of values of $\hat{\eta}_i$, with $i = 1, 2, ..., 25$. The next question to face is how one uses such a sample of discrete values to find the continuous probability density function itself [in this instance, the PDF in Fig. 3.3]. One must remember that the probability for η to have any given value, *exactly*, is zero. As shown by Eqn. (3.21), what the PDF does is to enable specification of the probability that η has a value in the infinitesimal *interval* $d\eta$ at η. Therefore, we should group our sample of values $\hat{\eta}_i$

into a number of discrete intervals, or "bins", that span their possible range, $0 \le \hat{\eta}_i < 1$. We can plot the relative number of chords per bin width for each bin as a histogram to represent the probability density function approximately. Using 10 bins of equal width $\Delta\hat{\eta}_i = 0.1$ for the sample of 25 chords, we obtain the histogram shown in Fig. 3.5. (The bins need not be of equal size.) The shaded portion under the histogram has unit area, as required of a PDF. Obviously, the sample we have used is too small to give a definitive picture of this function. However, we shall further explore what we have in order to illustrate several facets of Monte Carlo sampling.

Inspection of Fig. 3.5 shows that each chord in a given bin contributes a height of 0.4 to the value of $f(\hat{\eta}_i)$ in that bin. (The total "height" of all the bars is 10, which is 25×0.4). Thus, for example, the number of chords in the bin, $0.5 \le \hat{\eta}_i < 0.6$, is two. The poor representation of the PDF in the figure reflects the small number of chords in each bin. (No bin has more than five.) As will be seen shortly, the situation improves greatly with a sample much larger than 25. However, even with 25, using a smaller number of larger bins should reduce the statistical scatter somewhat. Figure 3.6 shows the same data, plotted now in five bins of width $\Delta\hat{\eta}_i = 0.2$. The new presentation of the same sample of chords conveys a different impression of how the actual density function might look. Its monotonic character is now more evident. Drawing a smooth curve from the origin through the midpoints at the top of each bin would give a reasonably good approximation to the density function in Fig. 3.3.

Finally, Fig. 3.7 presents results obtained with a sample of 1,000 random, isotropic chords in a sphere. The histogram shows that the probability density function varies, at least approximately, in linear fashion with the chord length. A histogram for 10^5 chords in 100 uniform bins (not shown) gives a very close approximation to Fig. 3.3.

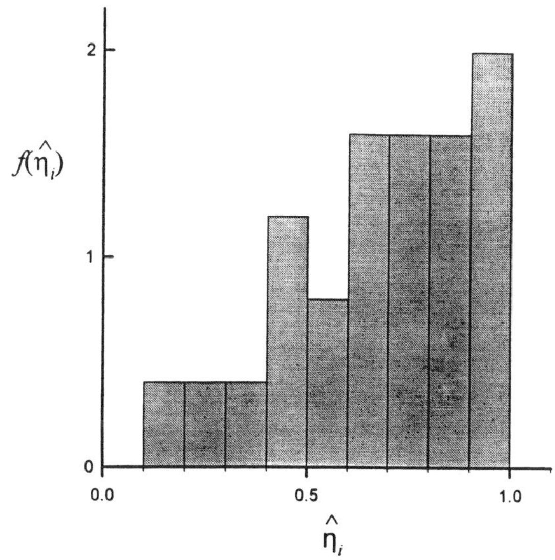

Fig. 3.5 Histogram from sample of 25 chords, showing rough approximation to PDF in Fig. 3.3. Data are grouped into ten bins of equal width. Chord lengths $\hat{\eta}_i$ are expressed in units of the sphere diameter.

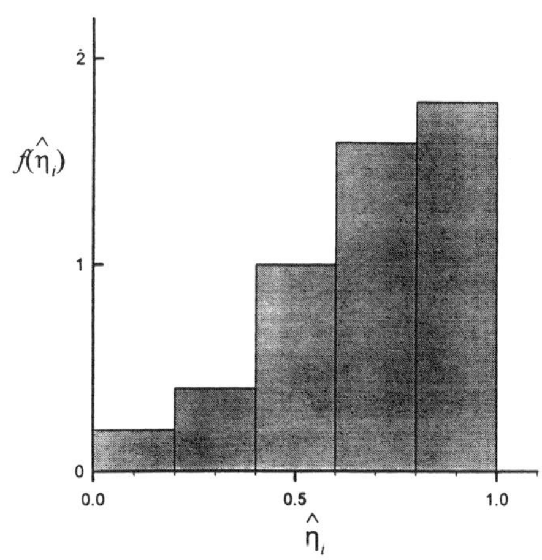

Fig. 3.6 Same data as in Fig. 3.5 for the 25 chords, grouped now in five bins of equal width.

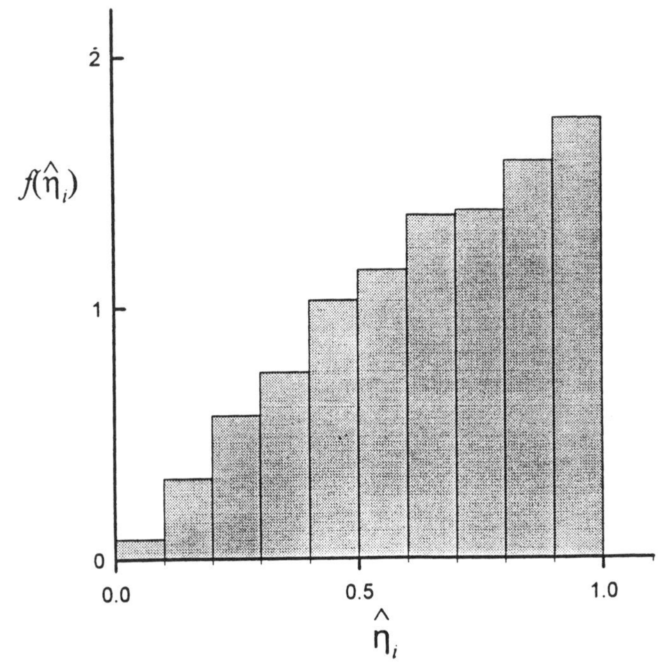

Fig. 3.7 Chord-length probability density function for a random sample of 1,000 chords.

Studies of chord lengths and LET unfolding have also been carried out for cylindrical cavities, this being a more natural geometry than spheres for gas proportional counters (Birkhoff et al. 1970; Turner et al. 1970; Borak 1994). Except for the sphere, very little exists in analytic form for isotropic (or other) chord lengths in other cavity shapes. The Monte Carlo method described here is ideally suited to provide numerical chord-length distributions for any shape. For cylinders, iterative Monte Carlo procedures were also employed to do the LET unfolding in a straightforward way. Most of this work was carried out for a "square" cylinder, *i.e.*, one having equal height and diameter. It is interesting that the mean isotropic chord length in the square cylinder is the same as that in its inscribed sphere. This rather remarkable circumstance follows from the Cauchy theorem (Cauchy 1908). The mean value $\bar{\eta}$ of uniform, isotropic chords that traverse any convex cavity is equal to four times the ratio of the volume V and surface area S of the cavity: $\bar{\eta} = 4V/S$. For the square cylinder and its inscribed sphere, it follows that $\bar{\eta} = 4R/3$, or two-thirds the diameter. This fact furnishes a useful check on the distributions generated in Monte Carlo studies. For instance, the mean for the 25 chords used for Fig. 3.5 turns out to be 0.6843, compared with the known population mean of exactly 2/3.

3.6 SOME MONTE CARLO STRATEGIES

In this section we briefly describe a few strategies and refinements that can be used to facilitate Monte Carlo computations.

3.6.1 Stratified Sampling

For the calculation of π in Fig. 3.1, random values of both x_i and y_i were generated over the interval [0,1). Statistical fluctuations in the results can be reduced if, instead of choosing both numbers at random, one takes a predetermined set of values uniformly spaced along the one of the axes. For example, a uniform sample of 1,000 equally spaced values, $0 \le x_i < 1$, is generated by the prescription,

$$x_i = 0.0005 + 0.001(i - 1) \quad , \quad i = 1, 2, ..., 1000 \tag{3.26}$$

For each x_i, the corresponding value of y_i can then be selected at random, and the calculation proceed as before. In principle, by making the x_i be uniform, this procedure should reduce the variance without biasing the computed statistical results. Similarly, in the paragraph following Eqn. (3.19) we described using a random number to make a choice between photoelectric absorption and Compton scattering in the computations. If we wish to generate, say, 1,000 photon histories in a Monte Carlo study, we can assign the first 884 collisions to be photoelectric absorption and the next 116 to be Compton scattering. Using these expected values eliminates statistical fluctuations in the choice of the collision type. Such pre-assignments are referred to as stratified sampling.

3.6.2 Importance Sampling

In some problems, relatively rare events can be of particular importance. A problem to be dealt with then is how to obtain a statistically significant sample of such events in a large background of other events that consume computer time. A good example is afforded by Monte Carlo calculations of neutron dose equivalent in soft tissue, which consists essentially of the elements, H, C, O, and N. A fast neutron interacts primarily with H and to a lesser extent with the three heavy nuclei. Some 93% of the first-collision absorbed dose from 1-MeV neutrons is due to elastic scattering from hydrogen . However, the 7% of the energy deposited by the heavy recoil nuclei struck by the neutrons becomes much more important when its contribution is weighted by the high-LET quality factors for these heavy recoils. Statistical sampling of these relatively infrequent, but important, collisions can be improved by means of the following scheme. When a collision occurs with a heavy nucleus, the otherwise single outcome can be replaced by a predetermined number n of independently computed outcomes, each given a statistical weight $1/n$. This method of importance sampling helps provide better statistical representations of rare events. Also, stratified sampling can be employed in this example to reduce the variance in the results by providing the expected number of neutron collisions with each element.

3.6.3 Smoothing Routines

In Fig. 3.5 we plotted the raw data obtained directly from the 25 random chords. One can reasonably expect the PDF to vary smoothly as a function of the chord length without up and down fluctuations. Thus, in random samples, there is often some correlation between the number of values of events that occur in a given bin and in its immediate neighbors. We saw in Fig. 3.6 that larger bins made the results look smoother. Another technique that is often useful in Monte Carlo work is to employ a smoothing routine to the raw data before plotting the histogram. As an example, the data from Fig. 3.5 were smoothed by arbitrarily replacing the ordinate of each bin by a weighted average. The average consists of 1/2 the original value for the bin plus 1/4 the sum of the ordinate values of the adjacent bins on either side. For instance, for the bin $0.4 \le \hat{\eta}_i < 0.5$ in Fig. 3.5, the original value $f(\hat{\eta}_i) = 1.2$ was replaced by the smoothed value,

$$\frac{1}{2}(1.2) + \frac{1}{4}(0.4 + 0.8) = 0.9 \tag{3.27}$$

For the two end bins, the replacement value was 3/4 of the original plus 1/4 that of the adjacent bin. The resultant smoothed histogram, displayed in Fig. 3.8, is much more informative than before. A variety of smoothing functions can be applied, depending on the problem of interest. The effect is akin to image enhancement.

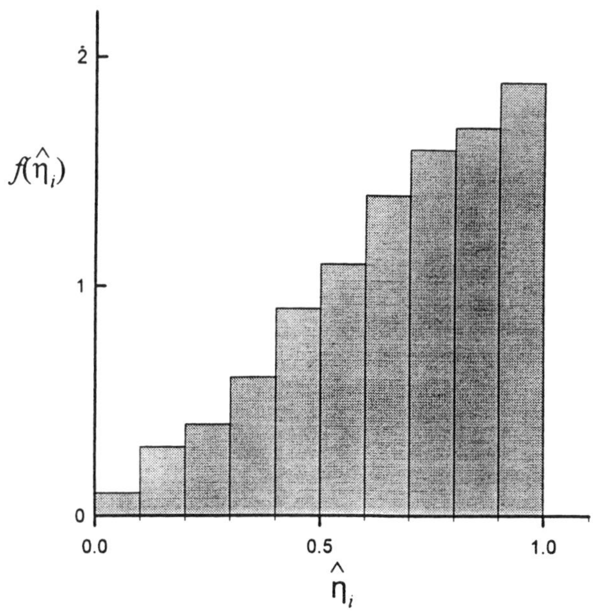

Fig. 3.8 Re-plot for the 25 chords from Fig. 3.5 after smoothing the raw data in adjacent bins, as described in text.

3.6.4 Splitting and Russian Roulette

This important technique is useful in deep-penetration problems, in which the sample of surviving radiation becomes increasingly small as particle transport is continued to greater depths. Only a small fraction of incident neutrons might get through a thick shield, providing a poor sample of the penetrating radiation needed for design of the shield. Most of the computer time in a "brute-force" Monte Carlo calculation might easily be spent in following particles that end up being absorbed before exiting the shield. To make the computations more efficient, one can designate a number of mathematical boundaries at fixed depths z_i in a shield and assign each incident neutron a statistical weight. When a neutron crosses a boundary at z_i, it is split into ν_i identical neutrons, each to be further transported, but with a reduced statistical weight, w_i/ν_i, where w_i was its weight as it crossed the boundary. Weight is preserved thereby, and the splitting gives ν_i as many independent histories from that boundary than otherwise.

Splitting becomes an even more effective way of sampling when, in addition, one applies Russian roulette. When a neutron being transported in the shield tends to wander back toward the surface through which it entered or it gets into some other "less desirable" region for the purpose of shielding determination, one can play Russian roulette with it. A random number is chosen to decide whether the neutron in question survives in the computations with some preassigned probability p, or whether it gets killed, with probability $1 - p$. If it survives, then its transport is continued, but with its weight increased by the factor $1/p$; otherwise, it is dropped from the calculations. Splitting with Russian roulette reduces the number of less desirable histories that would otherwise be continued in the computations.

3.6.5 Exponential Transformations

Exponential transformations provide another technique for dealing with deep-penetration problems. One artificially reduces the linear attenuation coefficient μ in regions of a target that are of less interest and increases it in regions that are important. These changes have the effect of reducing or increasing collision frequencies for purposes of efficient sampling. Appropriate weighting factors for the resultant particle-transport histories can be obtained in closed form.

3.7 ACKNOWLEDGMENT

I would like to thank James S. Bogard, Thomas B. Borak, and Darryl J. Downing for helpful comments and suggestions on this paper. I am also grateful to Dr. Bogard for preparing most of the figures.

Oak Ridge National Laboratory managed by Lockheed Martin Energy Research Corporation for the U.S. Department of Energy under contract DE-AC05-96OR22464.

3.8 REFERENCES

Atkinson, A.C. Tests of pseudo-random numbers. Appl. Stat. 29(2):164-177; 1980.

Bolch, W.E.; Turner, J.E.; Yoshida, H.; Jacobson, K.B.; Hamm, R.N.; Crawford, O.H. Product yields from irradiated glycylglycine in oxygen-free solutions: Monte Carlo simulations and comparison with experiments. Rad. Envir. Biophys. 37:157-166; 1998.

Borak, T. B. A method for computing random chord length distributions in geometrical objects. Rad. Res. 137:346-351;1994.

Carter, L. L.; Cashwell, E. D. Particle-transport simulation with the Monte Carlo method. TID-26607. Springfield, VA: Technical Information Service, U. S. Dept. of Commerce; 1975.

Cauchy, A. Oeuves completes, vol. 2. Paris, France: Gauthier-Villars; 1908.

Friedland, W.; Jacob, P.; Paretzke, H.G.; Merzagora, M.; Ottolenghi, A. Simulation of DNA fragment distributions after irradiations with photons. Rad. Envir. Biophys. 38:39-47; 1999.

Kalos, M.; Whitlock, P. Monte Carlo Methods. New York, NY: Wiley-Interscience; 1986.

Knuth, D. E. The art of computer programming, vol. 2: semi-numerical algorithms, 2nd ed. Reading, MA: Addison-Wesley; 1980.

Metropolis, N.; Ulam, S.M. The Monte Carlo method. J. Am. Stat. Assn. 44:335-341; 1949.

Newman, M. E. J.; Barkema, G. T. Monte Carlo methods in statistical physics. New York, NY: Oxford Univ. Press; 1999.

Rossi, H. H. Microscopic energy distribution in irradiated matter. In: Attix, F. H.; Roesch, W. C., eds. Radiation dosimetry. New York, NY: Academic Press; 1968.

Rossi, H. H.; Zaider, M. Microdosimetry and its applications. New York, NY: Springer; 1996.

Turner, J.E.; Wright, H.A.; Hamm, R.N. A Monte Carlo primer for health physicists. Health

Phys. 48:717-733; 1985.
Turner, J. E. Atoms, radiation, and radiation protection. New York, NY: John Wiley; 1995.
Ulam, S. M. Adventures of a mathematician. New York, NY: Charles Schribner's Sons; 1983.

3.9 AUTHOR PROFILE

Dr. Turner is a retired Corporate Fellow from Oak Ridge National Laboratory. He has served on the Board of Directors of the Health Physics Society and the editorial staffs of Health Physics and Radiation Research. He is a former member of the National Council on Radiation Protection and Measurements and a past President of the American Academy of Health Physics. Dr. Turner has authored numerous papers and three textbooks in radiation physics and dosimetry.

Chapter 4

UNCERTAINTY ANALYSIS

Thomas B. Kirchner

4.1 INTRODUCTION[1]

The goal of science is to uncover fundamental truths about the nature of the universe in the face of uncertainty. Although Smith and Ye (1998) claim that uncertainty analysis may be viewed as a statistical procedure, most practitioners of uncertainty analysis would probably conclude the converse, that statistics is one procedure used in uncertainty analysis. For example, uncertainty analysis must often deal with issues beyond the scope of statistics, e.g. a paucity or lack of data, complex dynamic models, extrapolation in spatial and temporal domains, decision-making involving societal concerns and regulatory issues, and multiple, alternative hypotheses. Nevertheless, statistics is an important tool for supporting decision-making in most scientific endeavors and many of the concepts for uncertainty analysis are derived from "classical" and "Bayesian" statistics.

Classical or "frequentist" statistics is strongly oriented toward describing a response in terms of a model. These models are generally restricted to equations having additive terms with linear, fixed parameters, (e.g. $a+bx+cx^2+dy+...$). An "error" term is added to represent the variability in the data not explained by the model. Bayesian statistics employs similar models, but assumes that the parameters of a model can be random factors. The Bayesian approach allows one to incorporate prior knowledge about the parameters into a statistical analysis, whereas the classical approach derives estimates of the parameters entirely from the data. These differences may seem small on the surface, but the use of Bayesian methodology in science, particularly the use of prior knowledge, is a source of considerable controversy (e.g. Mayo 1996).

Not all models used in science can be formulated as simple, linear response functions. Simulation models, based on dynamical representations of the functioning of systems, are also important. In a simulation model, a system is typically conceptualized

[1] The material in the chapter draws heavily upon Kirchner, T.B. In press. Estimation and application of uncertainty in asessment models. In Till, J.E. and H.R. Meyer. Risk Analysis for Radionuclides Released to the Environment, A Textbook on Environmental Risk Analysis for Radionuclides. Oxford University Press.

as a series of compartments linked by flows of material or energy. The compartments represent one or more components of the real system, such as the biomass of plants on a square meter of land surface or the amount of a radionuclide in the tissues of a plant. The rate of flow between compartments is often represented using differential equations, but some processes may be represented by difference equations or discrete events. For example, tillage of soil or the harvest of a crop can be represented as instantaneous events. Typical models of radionuclide transport are often based on the assumption of first order kinetics. If such models are constrained to a few compartments and inputs can be assumed to be continuous at a constant rate or to be a discrete, single exposure then analytical solutions of the rate equations can sometimes be derived. Many screening models, such as those of the NCRP (NCRP 1996), are derived in this way. However, more complex models or input scenarios require the iterative solution of the rate equations and discrete events over the specified time interval. The length of the time step for each iteration is determined by the type of equations being solved and, with differential equations, the accuracy with which the equations are to be solved. The applicability of a simulation model, its sensitivity to required input data, the quality of its predictions and the uncertainty surrounding its predictions must be based upon extensive validation, sensitivity and uncertainty analyses.

Quantitative uncertainty analysis, particularly in human and ecological risk assessment, remains controversial. Warren-Hicks and Moore (1998) discuss some of the benefits and criticisms of uncertainty analysis as applied to risk assessment. The potential benefits include:

- Quantitative uncertainty analysis improves transparency
- Quantitative uncertainty analysis improves credibility
- Quantitative uncertainty analysis focuses data collection
- Quantitative uncertainty analysis avoids worst-case assumptions
- Quantitative uncertainty analysis improves decision support

Potential criticisms include:

- Quantitative uncertainty analysis can be used incorrectly
- Quantitative uncertainty analysis cannot consider all sources of uncertainty
- Quantitative uncertainty analysis can be difficult to do well
- Quantitative uncertainty analysis is not conservative
- Quantitative uncertainty analysis is not always practical
- Quantitative uncertainty analysis is not accepted by everyone

- Quantitative uncertainty analysis can make decision-making difficult
- Quantitative uncertainty analysis can postpone decision-making

The validity of the benefits and criticisms are obviously dependent on specific situations. The goal of this chapter is to provide an introduction to uncertainty analysis that will help scientists to better understand the estimation and interpretation of uncertainty, and thus provide information with which to better evaluate the potential benefits and problems of uncertainty analysis. This chapter also focuses on issues of uncertainty in parameters of models, although some of the other issues in uncertainty analysis will be briefly discussed.

4.2 CLASSIFYING UNCERTAINTY

Uncertainty has been classified as framing, model, decision-theoretic and parameter (Shrader-Frechette 1996). Framing uncertainty arises early in the scientific process when deriving alternative hypotheses from a conceptual model of a system. Traditionally, science has opted for selecting a 2-value (true, not-true) frame for hypothesis testing, even in the face of high levels of uncertainty. The goal of seeking the truth (accuracy) leads scientists to focus on minimizing Type I errors (rejecting a null hypothesis when it is true), often without regard to Type II errors (not rejecting a null hypothesis when it is false) (Eberhardt and Thomas 1991). Type I errors are also called "false positive" errors, and Type II errors are called "false negative" errors (See Section 1.3.4). Scientists focus on Type I errors because it is generally considered more acceptable to reject a hypothesis that is true than to fail to reject a hypothesis that is false when faced with uncertainty. In addition, the probability of Type I errors can often be evaluated objectively, whereas Type II errors must often be evaluated subjectively (Zar 1984). In contrast, risk managers often are concerned with minimizing the chance that an assessment will falsely conclude that there is no significant risk, i.e. with minimizing Type II errors (Eberhardt and Thomas 1991). Risk assessors must often try to satisfy the proponents of both viewpoints when extrapolating from experimentally determined uncertainties to the environmental uncertainties required for projecting potential risks. For example, scientific credibility requires that dose estimates be as accurate as possible, whereas public acceptability of a study depends on establishing confidence that doses are not underestimated. Conservative dose estimates may be acceptable to the public, but be considered unacceptable to scientists because they are likely to result in false positive estimates, and unacceptable to risk managers because they may result in unnecessary destructive and costly cleanup activities.

Model uncertainty refers to the inherent nature of models as simplifications of real systems, thus subject to uncertainty as to whether the essential components and processes of a system are included in the model and whether those processes are adequately represented mathematically. Simplification often involves aggregation of variables, as when the human body is represented as a handful of compartments for the purpose of modeling exposure and dose. Such models can exhibit aggregation error (Gardner et al. 1982, O'Neill and Rust 1979, Mosleh and Bier 1992). For example, suppose concentrations of a radionuclide were measured in milk from individual cows and that these data were to be used to estimate the dynamics and uncertainty of radionuclide concentrations in the pool of milk that is obtained from a dairy. The rate of decline in the concentration of the radionuclide in milk is controlled by the combination of physical decay and biological elimination. Assume that the dynamics of the radionuclide in milk can be well described using first-order kinetics, i.e.

$$\frac{dQ}{dt} = kQ \tag{4.1}$$

where Q is the level of the contaminant in milk and k, the rate of loss, is negative. The data of Sasser (1965) show this to be a reasonable description of the concentration of in ^{131}I milk. The solution to this differential equation is

$$Q_t = Q_0 e^{kt} \tag{4.2}$$

where Q_0 is the initial value of Q and Qt is the value at time t. Sasser (1965) measured half-times for the rate of loss in 6 cows ranging from 0.60 to 0.85 days. A simulation model was constructed to represent the decline in milk concentration from each of the 6 cows. This individual-based model also computed the average milk concentration across the 6 cows:

$$Q_a = \sum_{1}^{6} Q_i / 6 \tag{4.3}$$

A second model was constructed to represent the dynamics of a hypothetical "herd". The dynamics of the herd was assumed to be identical to the dynamics for an individual cow, i.e.

$$\frac{dQ_h}{dt} = k_h Q_h \tag{4.4}$$

Thus, the herd model had a single compartment for the radionuclide in the pool of milk and was assigned a rate of loss equal to the mean of the 6 observed rate constants. The logarithm of Q_h is a linear function of time whereas the logarithm of Q_a is a concave curve (Fig. 4.1). The difference between Q_a and Q_h is "model bias", the error introduced by using an approximation of the system being modeled. Given the analytical solution for the model and the assumption that k is a random variable, an approximation for the function for the mean of milk from individual cows can be written

$$\overline{Q}_t \approx Q_0 e^{\overline{k}t + \left(t^2\sigma^2\right)/2} \tag{4.5}$$

with the corresponding differential equation

$$\frac{d}{dt}\overline{Q} = \overline{Q}\left(\overline{k} + t\sigma^2\right);\ k < 0 \tag{4.6}$$

Thus the rate of decline in mean milk concentration is expected to decrease as time increases. The difference between Q_h and Q_t is analogous to aggregation error. True aggregation error would be a component of the differences between Q_h and the true pooled milk concentration and thus would include the aggregation error associated with representing individual cows as a single compartment.

The magnitude of the potential uncertainty arising from model simplification and mis-specification is hard to estimate. Validation analyses are used to help establish confidence in the ability of a model to simulate a system and to provide an estimate of the magnitude of differences between model projections and the real system. Thus, validation attempts to delineate the domain of applicability of a model (Kirchner and Whicker 1984). The domain of applicability of a model refers to the set of conditions under which a model can be assumed to adequately represent the system of interest. Shrader-Frechette

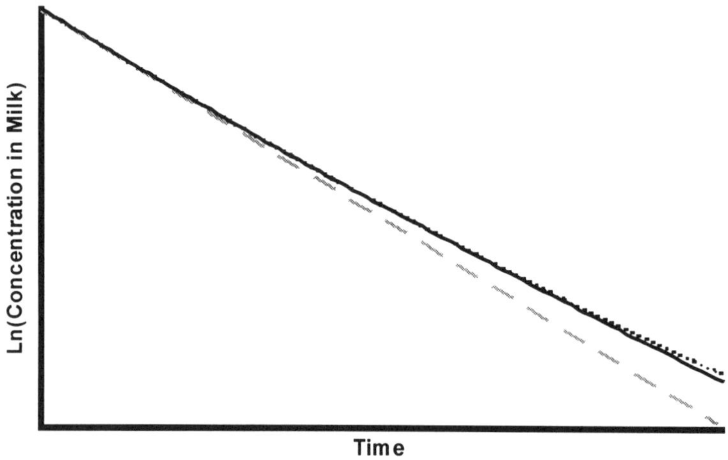

Fig. 4.1. Milk concentrations predicted from a one compartment model that uses the average of the rate constants from 6 cows (dashed line), from simulating the 6 cows and averaging the milk concentrations (solid line) and from a model for the mean that incorporates a time-dependent rate function

(1996) warns that model validation exercises which show favorable comparisons between model results and observations are often misinterpreted as an "affirmation of the consequent", in that the correspondences are interpreted as showing that the model is valid. In reality, such comparisons alone can be used to identify circumstances where a model is inadequate but never to show that it is a universally "valid" representation of the system. Naylor and Finger (1967) recommend using a three stage approach that promotes additional analyses to help put in perspective comparisons of model results to observations. Law and Kelton (1982) summarize the three step approach as:

- Establish the face validity for the model.
- Test the assumptions used in the model.
- Evaluate how well predictions used in the model correspond to observations from the system.

The first two steps in this scheme of validation can help to establish the credibility of a model by providing information to estimate the potential domain of applicability for the model even in the absence of quantitative data. Face validity of a model can be established through the review by experts in the field of study of the structure of the model, the form of the equations used in the model, and the values of parameters used in the model. Establishing the face validity of a model is important even when extensive quantitative evaluation of a model's performance can be made because the potential for aggregation error can be thus exposed and because differences can be masked by compensatory errors (Peterson and Kirchner 1998). Understanding the formulation of a model is necessary to evaluate its potential limitations due to inherent constraints or assumptions. When possible, assumptions should be tested empirically. It can also be useful to examine the effects of substituting alternative assumptions about mechanisms or structures. Such tests contribute to the understanding of the uncertainty in the model due to structural complexity.

In risk assessments involving projections well into the future, concerns about model uncertainty often overwhelm any concerns about parameter uncertainty (Casman et al. 1999). Whereas complex models may be appropriate for short-term projections, their effectiveness in modeling the distant future is usually questioned because of an expectation that one or more assumptions will become invalid. Simpler models with less restrictive assumptions may be appropriate in such cases. Casman et al. (1999) suggest that multiple models be used in making long term assessments, with rules being used to switch between the various models as they pass through their domains of applicability. Obviously, the scope and nature of the uncertainty will change through time, with a transition from probabilistic assessments to simple bounding of possible outcomes.

Decision-theoretic uncertainty is associated with the problem of making a decision in the absence of perfect knowledge, such as decisions involving alternative sources of risk. A typical source of uncertainty in decision-making related to human risks is whether to address risks and benefits at the societal or the individual level (Shrader-Frechette 1996). The choice between these frames of reference can have significant impacts on the decision. Treating a population as if it were composed of biological "equals" does not necessarily mean that a decision will impact all people equally.

Parameter uncertainty involves both natural variability in the parameters of a model and uncertainty arising from lack of knowledge. *Parameter* in this context refers to initial values of state variables, rate variables, and time dependent inputs to the model, sometimes called driving variables. Such uncertainty is often large, but the methods for propagating parametric uncertainty are well established. *Output* variable refers to a state variable or function of state variables whose response is of particular interest. Uncertainty in model predictions may change through simulation time or across the domain of applicability for a model (Breshears et al. 1989). The level of uncertainty in predictions may also change as the quality of the input data for a model changes.

Parameter uncertainty has been classified as aleatory and epistemic, or Types A and B uncertainty, respectively (IAEA 1989, ANSI 1997). Type A uncertainty is regarded as natural variability, which can be better characterized through additional research but cannot be reduced. Probability distributions associated with Type A uncertainty can be constructed from frequency distributions of data and are therefore distributions of objective probability. Type B uncertainty is introduced when insufficient knowledge about the exact value of a parameter leads one to assign a distribution to that parameter. The distribution represents one's confidence about the true value of the parameter. A distribution of Type B uncertainty is called a distribution of subjective probability because it reflects one's degree of belief about the likelihood of a parameter taking on various values. The distribution of a model output that arises from propagating only Type B uncertainty in model inputs should be considered a distribution of confidence regarding potential systematic error in the result and not as potential variability in the result. Smith and Ye (1998) suggest that Type A and B uncertainties are analogous to within-treatment and between-treatment variability, respectively, in statistical models. However, the error associated with Type B uncertainty is better described as systematic error because a difference between an assumed value of a constant but unknown parameter and its true value will yield a biased result. Understanding the distinction between true variability and distributions of confidence in uncertainty analysis is essential when interpreting uncertainties (Hattis and Burmaster 1994).

Whether the uncertainty assigned to a parameter is classified as Type A or B often depends on the endpoint of the analysis. For example, uncertainty about the mass of a specific individual would be of Type B, whereas variability in mass among individuals in a population would be of Type A. Data on the range of masses among individuals could be used to estimate the uncertainty in the inhalation rate of a particular individual, but the mass of that particular individual is truly a constant value. The distribution of masses is thus used to derive a distribution of confidence about the mass of a particular individual. Type B uncertainty can be derived using more subjective methods as well, such as by using the opinions of one or more experts. Formal elicitation methods are often recommended for constructing subjective distributions from several experts, although the cost for implementing such techniques can be high (Hoffman and Kaplan

1999). It is often difficult to partition uncertainty into Types A and B. For example, consider estimating the collective dose to a population. Data from a sample of a population can be used to estimate the parameters of the distribution of the population, such as its mean and variance. The true distribution of the population, if known, would contribute only Type A uncertainty to the estimate of collective dose. However, the distribution derived from the sample includes Type B uncertainty, represented by the distributions of confidence on the mean and variance.

In the past, it was often considered unnecessary to distinguish Type A and B uncertainties when doing an analysis because the same methods are used to propagate the uncertainties through models. Indeed, estimates of variability based on sparse data were often subjectively adjusted to account for the additional uncertainties associated with using the data in a different context. The goal in assigning subjectively derived distributions to parameters thought to be truly constant, and to subjectively adjusting distributions representing natural variability, was to increase confidence that decisions made using the model results would have a low likelihood of Type II (false negative) errors. However, there is growing interest among scientists and decision makers to produce realistic rather than conservative estimates of risk, to specifically identify the components of confidence versus expected frequency in the results from risk assessments, and to partition the contributions from subjective and objective uncertainties (Helton 1994, Hoffman and Hammonds 1994, MacIntosh et al. 1994). It is common for uncertainty due to lack of knowledge to dominate risk assessment analyses.

Some simulation models incorporate stochastic processes in their design. Stochastic processes are defined in terms of probabilities and are implemented using pseudorandom number generators (Naylor et al. 1968). Stochastic processes represent variability in factors such as consumption rates, precipitation or the growth rate of crops. Models that make use of stochastic processes need to be run many times in order to generate many sets of possible results, or "realizations". Summaries, such as the mean and variance of a compartment across realizations, are used as the output variables of the stochastic model. Analysis of the behavior of the model often requires that the output from each realization, frequently at each time step, be saved for examination after the completion of the simulation.

Whether a parameter is constant through time or variable through time can have a dramatic effect on the results. For example, assume a human is ingesting milk that has been contaminated with a radionuclide. The concentration of the radionuclide in the milk decays as a first order process due to radioactive decay and physical processes which remove the contamination from the pasture being consumed by the dairy cows. Loss from the human is also a first order process due to radioactive decay and biological elimination. The model for this system is thus

$$\frac{dQ_1}{dt} = -K_1 Q_1 \tag{4.7}$$

$$\frac{dQ_2}{dt} = IQ_1 - K_2 Q_2 \tag{4.8}$$

where Q_1 and Q_2 are the milk and human compartments, respectively, K_1 and K_2 are the loss rates and I is the ingestion rate of milk by the human. If the parameters for the model are assumed to be constant through time but are assigned distributions, representing either Type B uncertainty or pasture and animal specific variability, then the results are quite different than if the parameters are assumed to be stochastically variable through time (Fig. 4.2).

4.3 PROBABILITY DISTRIBUTIONS

The formulation of a quantitative representation of the uncertainty is a basic requirement for performing a quantitative parameter uncertainty analysis of a model. A minimum requirement is to define a possible range for each of the relevant parameters. Often one has information that indicates that there is a greater chance that a parameter will take on one value as compared to another. This information can be expressed by defining a probability distribution for the parameter. Subjective opinion or theoretical considerations are sometimes used to weight the selection of parameter values toward what are considered likely values. Often sparse data are combined with subjective knowledge to derive distributions for parameters. When there are sufficient data one can tabulate the frequency with which particular values of a parameter are likely to be observed and then fit a standard statistical distribution to the data.

4.3.1 Describing Distributions

Various ways for describing distributions are described in Section 1.2.6. Probability density functions (PDFs) and cumulative frequency distributions (CDFs) provide useful information about distributions (Figures 1.1 and 1.2). Properties of a distribution that often are of interest are its central tendency (mode, mean, or median),

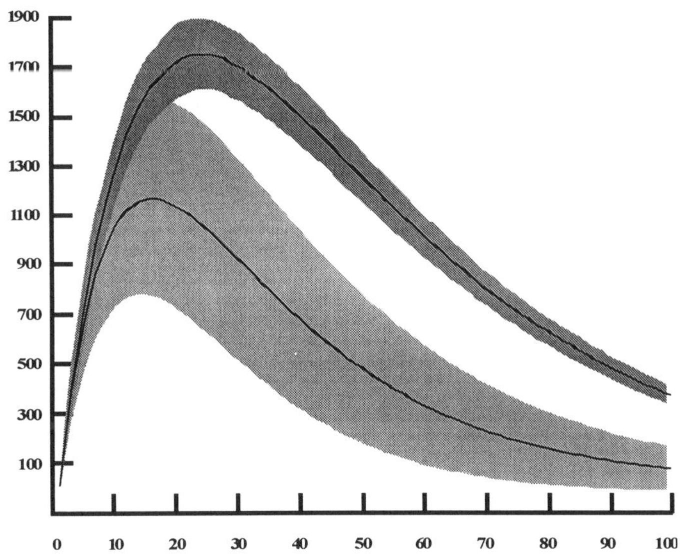

Fig. 4.2. Assuming parameters are stochastic but constant through time yields a wider interval of uncertainty than assuming parameters are stochastically variable through time. The means ± 1 standard deviation are plotted.

and a measure of its dispersion or spread (standard deviation or variance) (Section 1.2.6). In addition, measures of symmetry and shape (skewness and kurtosis) can be important, especially when one is interested in the tails of the distribution rather than its central tendency. The third central moment gives information about the asymmetry or skewness of a distribution:

$$\int (x-\mu)^3 f(x)dx \tag{4.9}$$

for continuous distributions and

$$\sum (x_i-\mu)^3 f(x_i) \tag{4.10}$$

for discrete distributions. Positive values of skewness indicate that the right tail of the distribution stretched to the right or, equivalently, that the mode of the distribution is shifted left (Fig. 4.3). The left tail of a distribution having negative skewness is stretched to the left. The fourth central moment provides information about the flatness of the distribution (Fig. 4.4).

$$\int (x-\mu)^4 f(x)dx \qquad (4.11)$$

$$\sum (x_i-\mu)^4 f(x_i) \qquad (4.12)$$

Percentiles or quantiles are also used to describe distributions. The γth quantile of a distribution is a value, X, for which the probability of observing values less than X is γ. The percentile of a distribution is equivalent to the quantile represented as a percentage. Thus 95% of all values are expected to be less than the 0.95 quantile. Percentiles are useful for representing the likely range of a parameter.

4.3.2 Types of Distributions

The type of a distribution is characterized by the functional form of its CDF or PDF. Although there are many types of distributions, only a few are commonly used in uncertainty analyses (Section 1.2.10 describes many of these). Normal and lognormal distributions are prevalent in analyses and arise because of the Central Limit Theorem (Section 1.2.12). An important property of lognormal distributions to remember is that the arithmetic mean lies to the right of the mode. If the geometric standard deviation exceeds 3.28 then the arithmetic mean will be larger than the 95th percentile for the distribution (Edelmann and Burmaster 1997). Poisson distributions are also frequently encountered in uncertainty analyses because they arise from such situations as events that occur randomly through time with a constant average rate, such as the decay of radioactive elements (Section 1.2.11.3).

A few distributions, particularly the uniform (Section 1.2.10.1), triangular (Section 1.2.9.2), trapezoidal and empirical distributions, are used frequently to represent uncertainty when data are scarce. The beta distribution (Section 1.2.10.5) is also used in

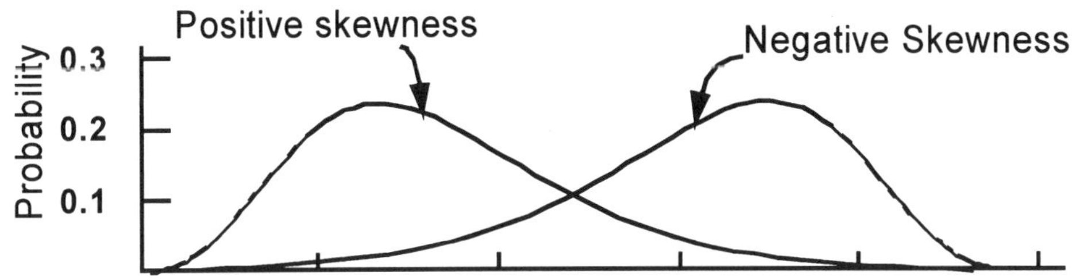

Fig. 4.3. Skewness of a distribution is a measure of asymmetry.

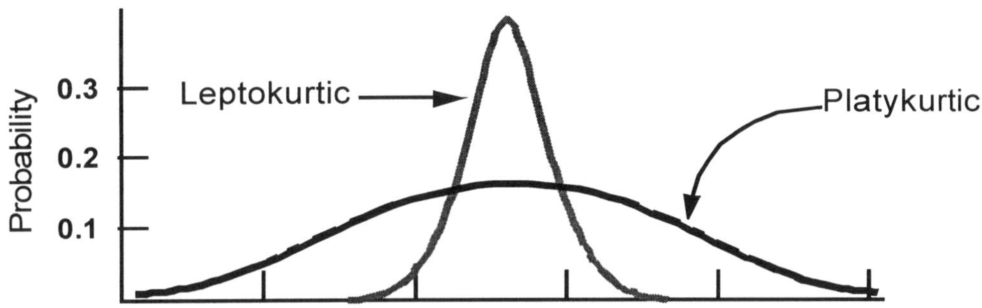

Fig. 4.4. Kurtosis describes whether a distribution is relatively flat (platykurtic) or pointed (leptokurtic).

these circumstances because it can be parameterized to take on a variety of shapes, extending from a uniform distribution to a mono-modal peaked distribution. Log-uniform, log-triangular and log-trapezoid distributions are often used when the possible range of a parameter is large.

Multivariate distributions (Section 1.2.13) represent conditional relationships among random variables. Such relationships can have large impacts on an analysis (Smith et al. 1992). Most of the univariate distributions do not have multivariate analogs. For example, there is no multivariate triangular distribution. Multivariate normal and lognormal distributions do exist and are often used when correlations exist between parameters. Multivariate lognormal distributions have a multivariate normal distribution of the logarithms of the variates. However, the correlation between the logarithms of the variates is not equal to the correlation between the variates (IAEA 1989). The relationship between the correlation coefficient for the logarithms, ρ', and the correlation coefficient for the untransformed variates, ρ, is given by

$$\rho' = \frac{\ln\left\{1 + \rho\sqrt{\left[\exp(\sigma_1^{\,2}) - 1\right]\exp(\sigma_2^{\,2}) - 1}\right\}}{\sigma_1 \sigma_2} \tag{4.13}$$

Iman and Conover (1982) have also developed a Monte Carlo sampling scheme for approximating the correlations between parameters having almost any marginal distribution. This method returns vectors of random numbers having specified ranked correlations.

4.3.3 Selecting Distributions for Inputs

Developing the distributions of uncertainty associated with the inputs to a model is usually a significant challenge. It is particularly important to specify exactly what the uncertainty on a parameter represents (Suter 1998). This definition must be consistent with the use of the parameter. For example, suppose a model uses a transfer coefficient to estimate the concentration of a contaminant in fruit given the concentration of the contaminant in the soil. In assigning uncertainty to the transfer coefficient one would need to consider whether the concentration in the fruit was expected to represent the concentration in fruit coming from a market that obtains fruit from several sources, the average concentration in several pieces of fruit, or the concentration in a single fruit from a backyard tree. Natural variability among trees and variability in the spatial distribution of the contaminant would be likely to cause one to assign a higher uncertainty to the transfer coefficient when applied to estimating contaminant concentrations from a single tree than when applied to estimating average concentrations from a market. The problem of properly specifying uncertainty so that a meaningful probability distribution can be assigned has led R.A. Howard and J. E. Matheson (1984) to suggest the clarity test as a conceptual means of ensuring proper specificity. The test involves the notion that, given sufficient resources, one could measure the parameter of interest. A properly specified parameter would thus give sufficient details to identify the conditions under which an experiment could be conducted to obtain the required data.

If one has sufficient data for an input parameter then the parameters for an appropriate distribution can often be estimated using standard statistical techniques of moment matching, maximum likelihood estimation or least squares minimization (Section 1.3). However, estimating the parameters of a distribution is only the first step in considering the uncertainty associated with the data. In addition, other uncertainties that could influence the interpretation of the data should be evaluated. For example, systematic errors are not identified by the replicated sampling used in most experiments but could possibly be estimated by experts. In small samples, low frequency events may have never been observed but could be accounted for by choosing an appropriate distribution. Although these considerations are not strictly objective, they may be necessary to account for factors that are assumed to be unimportant or non-existent in traditional statistical analyses.

One also needs to remember that the parameters derived from the analysis of samples are only estimates of the true parameters of the distribution that was sampled. The uncertainty in the distributional parameters is often expressed as confidence intervals on the parameters (e.g. Section 1.3.5). The confidence intervals placed on sample means and sample standard deviations are not truly objective measurements of variability, but simply represent a function of one's confidence about the likelihood that values around the sample parameter are equal to the parameter value of the true distribution. For example, the method used to construct the 95% confidence interval for a sample mean is expected to yield intervals that contain the true population mean in 95% of the cases where it is applied. Strictly speaking, in classical statistics a specific confidence interval does not enclose the true mean with a probability of 0.95; the true mean is either within that confidence interval or not. It is also not a statement that says that, if one were to collect more samples and compute the means of those samples, 95% of those sample means would be expected to fall within this particular interval.

Poisson distributions are frequently encountered in the fields of epidemiology and radiation measurement (Section 1.2.11.3). In the Poisson distribution, a is equal to both the mean and the variance of the distribution. If one observes n events where the distribution of events is Poisson then the $100(1-\alpha)\%$ confidence interval for a can be estimated by finding the smallest and largest a for which n events are expected with probabilities $1-\alpha/2$ and $\alpha/2$, respectively (Bailar and Ederer 1964) (Fig. 4.5). For example, if the number of observations is 20, then Poisson distributions with a as small as 13 and as large as 30.89 have the probability 0.025 or greater for observing 20 events (Table 4.1). The lower and upper confidence limits for a when there are n events can be found by searching for a values that satisfy

$$p = \sum_{x=0}^{n} \frac{e^{-a} a^{x}}{x!} \tag{4.14}$$

for p = 1- α /2 and p = α /2, respectively. Table 4.2 lists confidence limits for a given an observed n. Note that this method for finding confidence intervals gives slightly different answers than the method described in Section 1.3.10.3. The intervals are asymmetric and reflect the fact that the variance of the distribution changes in proportion to the mean.

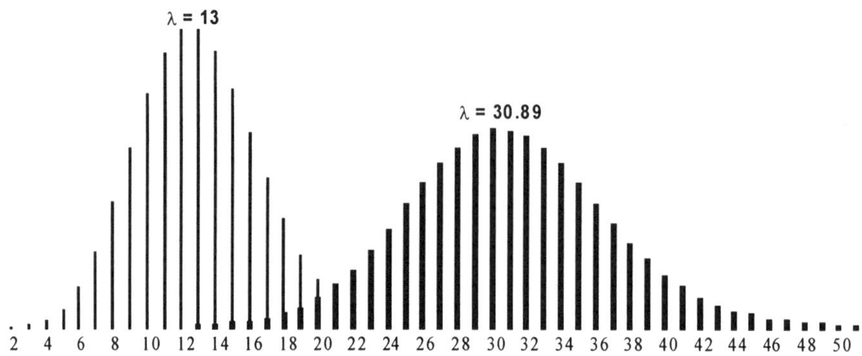

Fig. 4.5. A count of 20 could result from sampling any Poisson distribution with a mean between 13 and 30.89 with 95% confidence. This confidence interval is asymmetric.

Table 4.1. Confidence intervals for the Poisson distribution's single parameter a. For an observation of 20 counts, the 95% confidence interval for a is 13 to 30.89.

	Poisson, $a = 13$		Poisson, $a = 30.89$			Poisson, $a = 13$		Poisson, $a = 30.89$	
n	Frequency	Cumulative frequency	Frequency	Cumulative frequency	n	Frequency	Cumulative frequency	Frequency	Cumulative frequency
2	0				23	0.004	0.996	0.027	0.087
3	0.001	0.001			24	0.002	0.998	0.035	0.123
4	0.003	0.004			25	0.001	0.999	0.044	0.166
5	0.007	0.011			26	0.001	1.	0.052	0.218
6	0.015	0.026			27			0.059	0.277
7	0.028	0.054			28			0.065	0.343
8	0.046	0.1			29			0.07	0.412
9	0.066	0.166			30			0.072	0.484
10	0.086	0.252			31			0.071	0.555
11	0.101	0.353			32			0.069	0.624
12	0.11	0.463			33			0.065	0.689
13	0.11	0.573	0	0	34			0.059	0.748
14	0.102	0.675	0	0.001	35			0.052	0.799
15	0.088	0.764	0.001	0.001	36			0.044	0.844
16	0.072	0.835	0.001	0.002	37			0.037	0.881
17	0.055	0.89	0.002	0.005	38			0.03	0.911
18	0.04	0.93	0.004	0.009	39			0.024	0.935
19	0.027	0.957	0.006	0.015	40			0.018	0.953
20	0.018	0.975	0.01	0.025	41			0.014	0.967
21	0.011	0.986	0.015	0.04	42			0.01	0.977

Table 4.2. Ninety-five percent confidence levels for *a* given the number of observed events, n.

N	Lower	Upper	*n*	Lower	Upper	*n*	Lower	Upper
1	.243	5.573	27	18.606	39.285	65	51.045	82.849
2	.619	7.226	28	19.422	40.469	70	55.452	88.442
3	1.09	8.769	29	20.241	41.65	75	59.88	94.015
4	1.624	10.243	30	21.063	42.828	80	64.326	99.568
5	2.202	11.67	31	21.888	44.004	85	68.789	105.105
6	2.815	13.061	32	22.716	45.176	90	73.268	110.627
7	3.454	14.424	33	23.546	46.346	95	77.761	116.134
8	4.116	15.765	34	24.379	47.513	100	82.267	121.628
9	4.796	17.086	35	25.214	48.678	110	91.314	132.581
10	5.492	18.392	36	26.052	49.841	120	100.403	143.492
11	6.201	19.684	37	26.892	51.001	130	109.529	154.366
12	6.922	20.963	38	27.733	52.159	140	118.688	165.207
13	7.654	22.232	39	28.577	53.316	150	127.877	176.019
14	8.396	23.491	40	29.423	54.47	160	137.092	186.804
15	9.146	24.742	41	30.27	55.623	170	146.331	197.565
16	9.904	25.984	42	31.12	56.773	180	155.591	208.304
17	10.668	27.22	43	31.971	57.922	190	164.872	219.023
18	11.44	28.449	44	32.824	59.069	200	174.172	229.723
19	12.217	29.672	45	33.678	60.215	250	220.907	282.989
20	13.	30.89	46	34.534	61.359	300	267.953	335.943
21	13.788	32.102	47	35.392	62.502	350	315.237	388.659
22	14.581	33.31	48	36.251	63.643	400	362.708	441.188
23	15.378	34.513	49	37.111	64.782	450	410.332	493.564
24	16.179	35.712	50	37.973	65.92	500	458.085	545.811
25	16.985	36.906	55	42.302	71.592			
26	17.794	38.098	60	46.661	77.233			

Occasionally data cannot be well categorized into a typical statistical distribution, such as when the PDF shows multiple modes. In such cases an empirical distribution can sometimes be used. The CDF of an empirical distribution is derived from the observed data. However, caution must be used when sampling empirical distributions because the completeness of the distribution is correlated with sample size. Extreme but infrequent values may not be observed, yet such values may have profound influences on model predictions. In addition, empirical distributions may suffer from bias inherent in the data collection procedure.

Sometimes data are so sparse that it makes no sense to use statistical methods to try to fit distributions to the data. In such cases one may need to arbitrarily choose a distribution for a variable. A triangular distribution is frequently used when the true distribution is unknown but the extremes and a mean or modal value of the distribution can be estimated (Hodak 1994) (Fig. 1.3). The triangular distribution does not have to be symmetric. The trapezoidal distribution is similar to the triangular but does not have the emphasis of a single modal value (Seiler and Alvarez 1996). When only the maximum and minimum can be estimated a uniform distribution is the least biased assumption. When uncertainties are large one may want to consider using log-triangular, log-trapezoidal and log-uniform distributions. In these distributions it is the logarithms of the data that are triangularly, trapezoidally or uniformly distributed.

The most controversial method of assigning distributions to model inputs is undoubtedly the use of expert elicitation to identify parameters and their uncertainties when data are scarce or unavailable. Critics question whether opinions, even those of experts, deserve consideration in a quantitative assessment. Certainly the use of expert opinion can lead to biases, over- or under-estimation of uncertainties, or other errors. However, the alternative to using expert opinion is often to either reject performing a quantitative assessment of risks or to ignore the uncertainty in some model inputs and thus overstate confidence in model results.

When eliciting information from scientific experts one should be aware of the inherent differences in goals that can exist between the scientific and the risk assessment communities. Scientific training tends to focus on an experimentalist view of statistics, in which one tests hypotheses about the means of treatments and controls, and variability is an inconvenience that one tries to minimize through careful experimental design. Furthermore, one must recognize that there are confidence intervals associated with published estimates of both means and variances that should not be ignored when using data for risk assessment. The critics suggest that elicited estimates of uncertainty should be made sufficiently large so as to represent the combination of systematic and random errors. However, while scientists can objectively assess measurement and random errors, the estimation of potential systematic error is generally considered subjective, speculative and inappropriate for most scientific publications. Hence risk assessors are cautioned that they may need to employ meta-analysis and formal elicitation methods when eliciting information from scientists.

There has been concern expressed in the risk assessment literature over the last few years that scientists generally tend to underestimate the true uncertainties when reporting their results (overconfidence bias) (e.g. Schlyakhter 1994; Henrion and Fischhoff 1986). Although it appears that scientists do tend to overestimate their confidence in their data, it also appears that the problem has been overstated for some studies and based on improper interpretation of published uncertainties in others. For example, it has been pointed out that over the years physicists' measurements of the speed of light have routinely fallen outside the confidence intervals from previous experiments (Henrion and Fischhoff 1986). In this case, the published values of uncertainty undoubtedly represented the precision of the measurements made using a particular methodology, and did not (and could not) represent the potential systematic error in the measurements. Similar conclusions drawn by Schlyakhter (1994) in at least one of his examples appear to due to the failure to take sample size into account.

A thorough review of the literature on expert elicitation and decision analysis is beyond the scope of this chapter. Reviews of this subject area can be found in Tversky and Kanheman (1974), Morgan and Henrion (1992), Kahnemann et al. (1982), Meyer and Booker (1991) and Wright and Ayton (1994). Although research into the psychology of statistical reasoning and decision making by people has identified many potential problems, it is also true that much research has been aimed at the general population rather than at scientists or experts in particular fields. Motivated people who were knowledgeable in the subject area of the questions tended to perform better than the general public in experiments designed to test for biases and the use of non-statistical decision-making rules (Beach and Braun 1994). Thus, the practitioner should be cautious when dealing with expert opinion, but should not automatically reject the use of expert elicitation.

Formal protocols for eliciting expert judgement can help to ensure that the potential for cognitive biases is considered during the elicitation process. One such protocol involves 10 steps (Hora and Iman 1989, USNRC 1990):

1. Selection of issues and associated parameters
2. Selection of experts
3. Preparation of background information
4. Training of experts in the elicitation method
5. Presentation of issues, questions, and background information to experts

6. Analysis of issues, questions and background information by experts
7. Discussion of analyses of issues and background information by experts
8. Elicitations of judgement from the experts
9. Consolidation (and possibly aggregation) of the elicitation results
10. Review and communication of the elicitations

These steps include the features of other formal elicitation protocols, such as the Stanford/SRI Assessment Protocol (see Morgan and Henrion, 1990 for a discussion of this and other protocols). The five stages of the Stanford/SRI protocol are motivation, structuring, conditioning, encoding and verifying. The motivational stage introduces the nature of the problem and the expectations for the experts, including a discussion of potential motivational bias. The structuring phase proceeds to establish unambiguous definitions for the information to be elicited. The conditioning phase attempts to illuminate the cognitive processes of the experts for the benefit of both the expert and the analyst. This phase would include discussions about the various cognitive biases that could affect the expert's responses. The encoding phase seeks to generate a subjective probability distribution using the expert's knowledge. The verifying phase involves discussing the results with the expert to check for consistency.

Time-varying stochastic parameters in models impose additional difficulties when assigning distributions. A distribution for a time-varying stochastic parameter must be sampled at various times as the simulation proceeds. The length of time over which the parameter remains constant must correspond to the time interval assumed when assigning a distribution to the parameter. For example, assume that the consumption of contaminated milk by an individual is being simulated and that the milk consumption rate is assumed to vary from day to day. Furthermore, assume that new values for the stochastic parameters are selected at the start of each time step. If the time step for solving the rate equations exceeds 1 day then the variability in consumption through time is likely to be exaggerated, whereas if the time step is less than a day then the variability is likely to be depressed (Fig. 4.6). Clearly, either the definition of the variability in time-varying parameters must be consistent with the design of the model or vice versa.

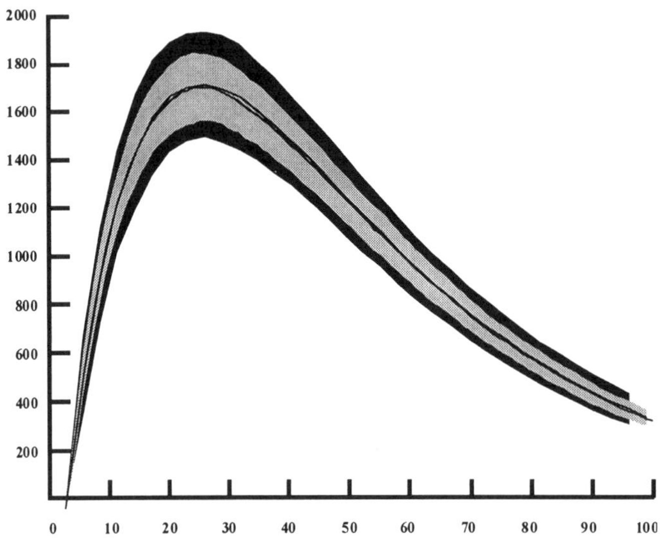

Fig. 4.6. Sampling every 3 time units produces a different estimate of uncertainty than sampling every time unit in this simple 2-compartment model.

4.4 METHODS OF PROPAGATION

Uncertainty can be propagated through models using analytical methods, mathematical approximation techniques, and Monte Carlo methods. Analytical methods and mathematical approximation techniques are usually only employed with relatively small models of limited complexity when there is no time dependent stochasticity. Discontinuities in model functions or the use of time dependent driving variables make the analytical and mathematical approximation techniques difficult or impossible to use.

4.4.1 Analytical Methods

Analytical methods to propagate uncertainty give exact solutions for the expectation and variance for the sum, difference and product of random variables but can only give estimates for quotients. These formulas do not require that one make assumptions about the types of distributions associated with the variables. However,

these formulas also do not provide information about the form of the resulting distribution. The formulas for these arithmetic operations are (Mood et al. 1974):
4.4.1.1 Sum and difference of random variables:

$$E\left[\sum_{i=1}^{n} X_i\right] = \sum_{i=1}^{n} E[X_i] \tag{4.15}$$

and

$$\operatorname{var}\left[\sum_{i=1}^{n} X_i\right] = \sum_{i=1}^{n} \operatorname{var}[X_i] + 2\sum_{i=1}^{n-1}\sum_{j=i+1}^{n} \operatorname{cov}(X_i, X_j) \tag{4.16}$$

for a sum and

$$\operatorname{var}\left[\sum_{i=1}^{n} X_i\right] = \sum_{i=1}^{n} \operatorname{var}[X_i] - 2\sum_{i=1}^{n-1}\sum_{j=i+1}^{n} \operatorname{cov}(X_i, X_j) \tag{4.17}$$

for a difference. If the variables in the sum or difference are *independent*, the formula for the variance reduces to:

$$\operatorname{var}\left[\sum_{i=1}^{n} X_i\right] = \sum_{i=1}^{n} \operatorname{var}[X_i] \tag{4.18}$$

4.4.1.2 Product of random variables.

$$E[XY] = \mu_X \mu_Y + \operatorname{cov}[X,Y] \tag{4.19}$$

and

$$\begin{aligned} \operatorname{var}[X,Y] &= \mu_Y^2 \operatorname{var}[X] + \mu_X^2 \operatorname{var}[Y] + 2\mu_X\mu_Y \operatorname{cov}[X,Y] - (\operatorname{cov}[X,Y]) \\ &+ E[(X-\mu_X)^2(Y-\mu_Y)^2] + 2\mu_Y E[(X-\mu_X)^2(Y-\mu_Y)] + 2\mu_X E[(X-\mu_x)(Y-\mu_Y)^2] \end{aligned} \tag{4.20}$$

If X and Y are *independent:*

$$E[X,Y] = \mu_X \mu_Y \tag{4.21}$$

and

$$\operatorname{var}[X,Y] = \mu_Y^2 \operatorname{var}[X] + \mu_X^2 \operatorname{var}[Y] + \operatorname{var}[X]\operatorname{var}[Y] \tag{4.22}$$

4.4.1.3 Quotient of random variables.

$$E\left[\frac{X}{Y}\right] \approx \frac{\mu_X}{\mu_Y} - \frac{1}{\mu_Y^2}\operatorname{cov}(X,Y) + \frac{\mu_X}{\mu_Y^3}\operatorname{var}[Y] \tag{4.23}$$

and

$$\operatorname{var}\left[\frac{X}{Y}\right] \approx \left(\frac{\mu_X}{\mu_Y}\right)^2 \left(\frac{\operatorname{var}[X]}{\mu_X^2} + \frac{\operatorname{var}[Y]}{\mu_Y^2} - \frac{2\operatorname{cov}[X,Y]}{\mu_X \mu_Y}\right) \tag{4.24}$$

If X and Y are *independent*:

$$E\left[\frac{X}{Y}\right] \approx \frac{\mu_X}{\mu_Y} + \frac{\mu_X}{\mu_Y^3}\operatorname{var}[Y] \tag{4.25}$$

and

$$\operatorname{var}\left[\frac{X}{Y}\right] \approx \left(\frac{\mu_X}{\mu_Y}\right)^2 \left(\frac{\operatorname{var}[X]}{\mu_X^2} + \frac{\operatorname{var}[Y]}{\mu_Y^2}\right) \tag{4.26}$$

4.4.2 Formulas for normal and lognormal distributions

Parameters are frequently assumed to have normal or lognormal distributions because experiments show that the distribution is reasonable, because of theoretical considerations, or because of computational considerations. The following theorems are useful in propagating uncertainties when dealing with normal or lognormal distributions.

4.4.3 Linear operations

Let X be a normally distributed random variable with mean μ_X and variance σ_X^2. This can be denoted

$$X \sim N\left(\mu_X, \sigma_X^2\right) \tag{4.27}$$

Linear combinations of normally distributed random variables are also normally distributed:

$$\sum_1^n a_i X_i \sim N\left(\sum_1^n a_i \mu_i, \sum_1^n a^2{}_i \sigma_i^2\right) \tag{4.28}$$

Adding a constant to normally distributed values shifts the mean but has no impact on the variance:

$$C + X \sim N\left(\mu_X + C, \sigma_x^2\right) \tag{4.29}$$

4.4.4 Geometric means and standard deviations

Assume X has a lognormal distribution. Then let *m* equal the mean of the logs of X:

$$m = \mu_{\ln(x)} \tag{4.30}$$

and let *v* equal the variance of the logs of X:

$$v = \sigma_{\ln(x)}^2 \tag{4.31}$$

The geometric mean for the distribution of X, GMX, is:

$$\mathrm{GM}_X = e^m = e^{\mu_{\ln(x)}} \tag{4.32}$$

and the geometric standard deviation, GSD_X, is

$$\mathrm{GSD}_X = e^{v^{1/2}} = e^{\sigma_{\ln(x)}} \tag{4.33}$$

To convert from geometric means and standard deviations to arithmetic means and standard deviations use the following formulas.

$$\mu_x = e^{\left\{\ln(\mathrm{GM}_X) + \frac{[\ln(\mathrm{GSD}_X)]^2}{2}\right\}} \tag{4.34}$$

$$\sigma_X^2 = e^{\left(2\left\{\ln(\mathrm{GM}_X)+[\ln(\mathrm{GSD}_X)]^2\right\}\right)} - e^{\left\{2\ln(\mathrm{GSD}_X)+[\ln(\mathrm{GSD}_X)]^2\right\}} \tag{4.35}$$

$$\mathrm{GM}_X = e^{\left[\ln(\mu_X) - \frac{1}{2}\ln\left(\frac{\sigma_X^2}{\mu_X^2}+1\right)\right]} = \frac{\mu_X}{\sqrt{\frac{\sigma_X^2}{\mu_X{}^2}+1}} \tag{4.36}$$

$$\mathrm{GSD}_X = e^{\sqrt{\ln\left(\frac{\sigma_x{}^2}{\mu_X{}^2}+1\right)}} \tag{4.37}$$

4.4.5 Propagation Using Interval Estimates

If the uncertainty is defined in terms of an interval, such as X±δX, then the formulas for propagating the uncertainties through sums, differences, quotients and products are similar to those for propagating means and variances (Taylor 1982).

4.4.5.1 Sum and difference. If several independent variables X, …, W are estimated with uncertainties expressed as the interval ±δX, …, ±δW, then the sum and difference of the variables

$$q = X + \ldots + Z - (U + \ldots + W) \tag{4.38}$$

would have the uncertainty

$$\delta q = \sqrt{(\delta X)^2 + \ldots + (\delta Z)^2 + (\delta U)^2 + \ldots + (\delta W)^2} \tag{4.39}$$

If the variables show complete correlation then the formula for propagation is simplified to:

$$\delta q = \delta X + \ldots + \delta Z + \delta U + \ldots \delta W \tag{4.40}$$

The assumption of independence should be examined carefully when using these methods (Sections 1.2.9 and 4.5). For example, if the errors on 3 measurements were 2, 5 and 3 then the error on the sum would be 6.2 if the measurements are independent but 10 if they are highly correlated.

4.4.5.2 Products and quotients. If several independent variables X, ..., W have associated with them small intervals of uncertainty ±δX, ..., ±δW then the uncertainty on a product or quotient of the variables

$$q = \frac{X \times \ldots \times Z}{U \times \ldots \times W} \tag{4.41}$$

is given by

$$\frac{\delta q}{|q|} = \sqrt{\left(\frac{\delta X}{|X|}\right)^2 + \ldots + \left(\frac{\delta Z}{|Z|}\right)^2 + \left(\frac{\delta U}{|U|}\right)^2 + \ldots + \left(\frac{\delta W}{|W|}\right)^2} \tag{4.42}$$

If the variables are completely correlated (ρ=1) then

$$\frac{\delta q}{|q|} = \frac{\delta X}{|X|} + \ldots + \frac{\delta Z}{|Z|} + \frac{\delta U}{|U|} + \ldots + \frac{\delta W}{|W|} \tag{4.43}$$

4.4.6 Mathematical Approximation Techniques

Approximations of the expected value and variance for a function can be computed from the means and variances of the parameters and the partial derivatives of the function at the mean values of the parameters. This method, sometimes called the Delta method, is based upon a Taylor series expansion of the functions.

4.4.6.1 Mean.

$$E[f(X_1,X_2,\ldots,X_n)] \approx f(\mu_1,\mu_2,\mu_3,\ldots,\mu_n) + \frac{1}{2}\sum_{j=1}^{n}\left[\frac{\partial^2 f}{\partial X^2{}_j}\right]\times \mathrm{var}[X_j] + \sum_{j=1}^{n}\sum_{j'>j}^{n}\left[\frac{\partial^2 f}{\partial X_j \partial X_{j'}}\right]\times \mathrm{cov}[X_j,X_{j'}] \quad (4.44)$$

If all of the variables are *independent* then

$$E[f(X_1,X_2,\ldots,X_n)] \approx f(\mu_1,\mu_2,\mu_3,\ldots,\mu_n) + \frac{1}{2}\sum_{j=1}^{n}\left[\frac{\partial^2 f}{\partial X^2{}_j}\right]\times \mathrm{var}[X_j] + \sum_{j=1}^{n}\sum_{j'>j}^{n}\left[\frac{\partial^2 f}{\partial X_j \partial X_{j'}}\right]\times \mathrm{cov}[X_j,X_{j'}] \quad (4.45)$$

$\frac{\partial^2 f}{\partial X^2{}_j}$ is the second derivative of f with respect to X_j evaluated at μ_j.

4.4.6.2 Variance.

$$\mathrm{var}[f(X_1,X_2,\ldots,X_n)] \approx \sum_{j=1}^{n}\left[\frac{\partial f}{\partial X_j}\right]^2 \times \mathrm{var}[X_j] + 2\sum_{j=1}^{n}\sum_{j'>j}^{n}\left[\frac{\partial}{\partial X_j}\right]\times \mathrm{cov}[X_j,X_{j'}] \quad (4.46)$$

If X_j and $X_{j'}$ are independent, then $\mathrm{cov}[X_j,X_{j'}] = 0$.

$$\text{var}[f(X_1, X_2, \ldots, X_n)] \approx \sum_{j=1}^{n} \left\{ \left[\frac{\partial f}{\partial X_j} \right]^2 \times \text{var}[X_j] \right\} \tag{4.47}$$

This method of approximation is often used to estimate the mean and variance of functions of random variable (Table 4.3).

4.4.6.3 Example. Consider the function e^Y where Y is a distribution having mean μ_Y and variance σ_Y^2. The partial derivatives are:

$$\frac{\partial e^Y}{\partial Y} = e^Y \tag{4.48}$$

$$\frac{\partial^2 e^Y}{\partial Y^2} = e^Y$$

so the estimators for the mean and variance are:

$$\mu_X = e^{\mu_Y} + \frac{1}{2} e^{\mu_Y} \sigma_Y^2 = e^{\mu_Y} \left(1 + \frac{1}{2} \sigma_Y^2 \right) \tag{4.49}$$

$$\sigma_X^2 = e^{2\mu_Y} \sigma_Y^2 \tag{4.50}$$

Compare these estimators with those in Table 4.3, which were derived by assuming that Y had a normal distribution.

Table 4.3. Approximations of the mean and variance for several frequently used mathematical functions [a].

Function	Mean	Variance
$x^* = e^y$	$\mu_x = e^{\mu_y + \frac{\sigma_y^2}{2}}$	$\sigma_x^2 = e^{2\mu_y + 2\sigma_y^2} - e^{2\mu_y + \sigma_y^2}$
$x = \ln(y)$	$\mu_x = \ln(\mu_y) - \frac{\sigma_y^2}{2\mu_y^2}$	$\sigma_x^2 = \frac{\sigma_y^2}{\mu_y^2}$
$x^{**} = \ln(y)$	$\mu_x = \ln(\mu_y) - \frac{1}{2}\ln\left(\frac{\sigma_y^2}{\mu_y^2} + 1\right)$	$\sigma_x^2 = \ln\left(\frac{\sigma_y^2}{\mu_y^2} + 1\right)$
$x = \sin(y)$	$\mu_x = \sin(\mu_y) \times \left(1 - \frac{\sigma_y^2}{2}\right)$	$\sigma_x^2 = \sigma_y^2 \times [\cos(\mu_x)]^2$
$x = \cos(y)$	$\mu_x = \cos(\mu_y) \times \left(1 - \frac{\sigma_y^2}{2}\right)$	$\sigma_x^2 = \sigma_y^2 \times [\sin(\mu_x)]^2$
$x = \sqrt{y}$	$\mu_x = \sqrt{\mu_y} - \frac{1}{8} \times \mu_y^{-\frac{3}{2}} \times \sigma_y^2$	$\sigma_x^2 = \frac{\sigma_y^2}{4\mu_y}$
$x = c^y$	$\mu_x = c^{\mu_y} + \frac{1}{2}\sigma_y^2 \ln(\mu_y)^2 c^{\mu_y}$	$\sigma_x^2 = \sigma_y^2 c^{2\mu_y} \ln(\mu_y)^2$
$x^* = c^y$	$\mu_x = e^{\ln(c)\mu_y + \frac{\ln(c)^2 \sigma_y^2}{2}}$	$\sigma_x^2 = e^{2\ln(c)\mu_y + 2\ln(c)^2\sigma_y^2} - e^{2\ln(c)\mu_y + \ln(c)^2\sigma_y^2}$
$x = y^c$	$\mu_x = \mu_y^c + \frac{1}{2}c(c-1)\mu_y^{c-2}\sigma_y^2$	$\sigma_x^2 = \sigma_y^2 c^2 \mu_y^{2(c-1)}$

[a] The equations for the expressions marked * or ** assume that the distribution for y is normal or lognormal, respectively, and give exact results for that case. The remaining equations are first order approximations derived using a Taylor's series.

4.5 COVARIANCE

Covariance is a measure of the linear relationship between two random variables (Section 1.2.9). One error that can be made when propagating uncertainties analytically is to ignore the covariance between two or more terms which results from having the same parameter in several terms. For example, assume that there is uncertainty in the amount of deposition on a farm following a release of radionuclides into the air. Logically, the transport of radionuclides through the soil-pasture-cow-human pathway should be independent of the transport through the soil-vegetable-human pathway. Nevertheless, if doses due to ingestion of milk and vegetables are computed independently with their uncertainties and then summed, then the uncertainty on the total ingestion dose will be underestimated if the milk dose is assumed to be independent of the vegetable dose. This situation arises because there should be a high correlation between the amounts of deposition assumed to occur on the pasture and on the garden even though, in reality, the amounts of deposition were single values. Algebraically, this can be represented

$$Y = A \times (X + Z) \tag{4.51}$$

where A, X, and Z are independent of one another. One way to compute the variance in Y would be

1. Apply the formula for a sum to compute the variance of X+Z. Since the variables are independent, one can ignore the covariance term.
2. Apply the formula for a product using var[A] and var[X+Y], also assuming independence.

However, notice what happens if the equation is rearranged:

$$Y = AX + AZ \tag{4.52}$$

It may seem reasonable to compute var[AX] and var[AY] using the formula for products and then apply the formula for a sum of independent variables to compute var[Y]. However, the terms AX and AY are no longer independent because they share the common factor A. The correct solution can be achieved by performing the operations as listed in steps 1 and 2.

4.6 MONTE CARLO METHODS

Monte Carlo methods (discussed in Chapter 3) are frequently used in uncertainty analysis because they can be applied to models of any size and with models having complex equations. However, Monte Carlo methods typically involve large numbers of simulations, and thus may be costly for large models. The Monte Carlo method involves choosing parameter or other input values using a random or stochastic selection scheme. Typically, parameters are assumed to remain constant throughout a simulation and are selected prior to the start of each simulation. Numerous simulations are run, and the values of the output variables for each of those simulations are saved for later analysis (Fig. 4.7). If the model is a dynamic simulation model then the values of the output variables may be saved at one or more times during the simulation. If parameters are assumed to vary stochastically through time, such as one might assume for the rate of ingestion of fish or wild game by an individual, then additional sampling must occur throughout the simulation.

Monte Carlo methods are generally easy to implement. However, there are circumstances which, should they arise, can lead to unpredictable results. One of these situations involves the use of distributions with infinite tails, such as the normal distribution. The presence of an infinite tail means it is possible, even if unlikely, that a value will be generated that is near, or even equal to, zero, that a negative value will be generated, or that a large positive or negative value will be produced. The likelihood of seeing of such low-probability events as the number of values generated becomes large. For example, for 10,000 values selected from a normal distribution there is only a 0.05 probability, with 95% confidence, of not seeing a value selected beyond $\pm 3.4\sigma$. If the parameter cannot logically take on such extreme values then errors might arise. For example, a negative loss rate or a near zero value in a denominator can easily produce undesired behavior. Truncation of the input distribution at known or logical limits can help to avoid this problem.

It is possible to specify input distributions that yield output distributions that can not be usefully characterized. For example, the ratio of 2 variables having standard normal distributions results in a Cauchy distribution. The Cauchy distribution has tails that never converge to 0. Hence, the distribution has no moments, i.e. no mean or variance. Sample mean and variances can be computed from samples taken from a Cauchy, or course, but the mean and variance are unlikely to converge as sample size increases. In fact, variance may increase as sample size increases.

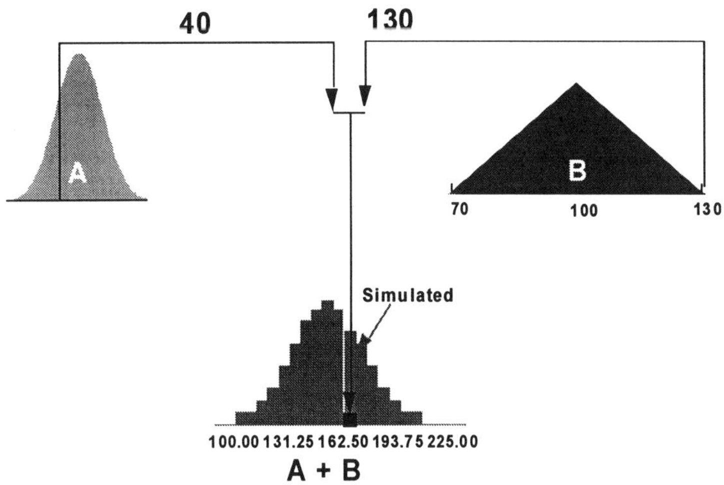

Fig. 4.7. Monte Carlo sampling involves selecting values from the distribution of each parameter and using them in the model equations to produce a result. This procedure is repeated many times to generate a distribution of results.

Many of the uniform random number generators available on computer systems are written for fast execution rather than for generating good sequences of numbers. The length of the period on one common generator, the linear congruential method, is required by the ANSI C standard to be only 32767. Autocorrelation is also a problem with many standard random number generators supplied with compiler libraries. Press et al. (1988) recommend a random number generator that uses shuffling to break up sequential correlations.

4.7 SAMPLING DESIGNS

An uncertainty analysis based upon Monte Carlo simulations requires that a sampling design be implemented. Simple random sampling is the least complicated design but is often also the least efficient. In simple random sampling the values of parameters are chosen completely at random from their distributions. In some instances parameters may be known to be correlated. Simulating correlated parameters is possible, but requires one either to use multivariate pseudo-random number generators or to introducing correlations by generating samples of independent values and then manipulating the ordering of the values within those samples (Iman and Conover 1982).

Stratified designs can significantly reduce the number of simulations required to estimate the mean and variance of output variables, which can be important when dealing with complicated models or when numerous scenarios to be considered (Section 3.6.1). Importance sampling (Section 3.6.2) is a highly stratified design that can be used to minimize the number of simulations by concentrating the sampling in those regions of the distributions of the input parameters that give rise to extreme values in the results (Clark 1961, Fishman 1996). Another stratified design, Latin Hypercube Sampling (LHS), stratifies the distribution of each parameter into n quantiles, each of which contains100/n percent of the distribution (McKay et al. 1979). Thus, it is equally probable that a value chosen at random from the distribution would occur in any particular quantile. In the LHS design each quantile of each simulation is sampled only once during n simulations. Values are chosen at random from within the range defined for a selected quantile.

As an example of a Latin Hypercube design, consider a Monte Carlo simulation that involves 5 parameters whose distributions are stratified into 10 quantiles (Table 4.4). The process starts by constructing a 5 by 10 matrix. The rows of the 5 by 10 matrix identify parameters and the columns are simulations. Numbers from 1 to 10 are assigned randomly to each cell within a row without replacement, so that a given number appears only once in a row. The numbers in the cells identify which quantile is to be sampled at the start of each simulation. For example, a 1 indicates that sampling is to be restricted to the quantile for 0 to 10%. Pseudo-random number generators are used to create a corresponding matrix of parameter values to be used in the simulations by selecting values at random from within the specified quantile. A variation of the Latin Hypercube design uses the midpoint of all quantiles rather than selecting a value at random from within the quantile, although this technique can bias the estimate of variance or introduce other errors in certain cases (Morgan and Henrion 1992). If more than 10 simulations are run then new matrices are constructed for each set of 10 simulations. However, it is usually the case that the number of quantiles is set equal to the total number of simulations to be run. To avoid bias, the simulations should not be terminated before the completion of an entire matrix. The Latin Hypercube sampling scheme has also been expanded to enable correlated sampling, even using marginal distributions of different types (Iman and Conover 1982).

Table 4.4. A Latin hypercube design for 5 parameters and 10 simulations. The numbers in the rows of the parameters show which quantile is sampled for the listed simulation. Thcsc valucs are assigned at random without replacement in each row.

Parameter	Simulation									
	1	2	3	4	5	6	7	8	9	10
A	4	7	1	10	8	2	5	9	3	6
B	1	8	2	4	6	3	10	5	7	9
C	7	3	6	9	2	5	8	1	4	10
D	5	4	3	6	9	1	2	7	10	8
E	9	6	2	5	1	4	7	8	10	3

Hoffman and Hammonds (1994) argue that, for some questions, it is important to distinguish uncertainty due to natural variability (Type A) from uncertainty due to lack of knowledge (Type B). To achieve such partitioning the sampling scheme must nest the sampling for the parameters associated with natural variability within a sampling scheme for the parameters associated with lack of knowledge. For example, assume parameters A, B and C have natural variability and parameters D, E and F have constant but unknown values. A set of, say, 1000 Monte Carlo simulations could be run selecting only values of A, B and C for each combination of values selected for D, E and F. If 1000 sets of values for D, E and F were selected then a total of 10^6 simulations would need to be run and 1000 "families" of distributions for the output variable would be generated (Fig. 4.8). Hoffman and Hammonds also point out that, in some cases, the means and variances for the distributions of A, B, and C can also be classified as constant but unknown values, such as when they are estimated from sample data, and assigned to the Type B category in the sampling scheme. This technique enables one to place confidence bounds around the CDF for the output distribution, where the uncertainty in the CDF arises from the lack of knowledge about the fixed parameters. The nested sampling design required for these analyses can be cumbersome and time consuming. Bogen (1995) suggests some other innovative approaches for simplifying the joint assessment of lack of knowledge uncertainty and variability.

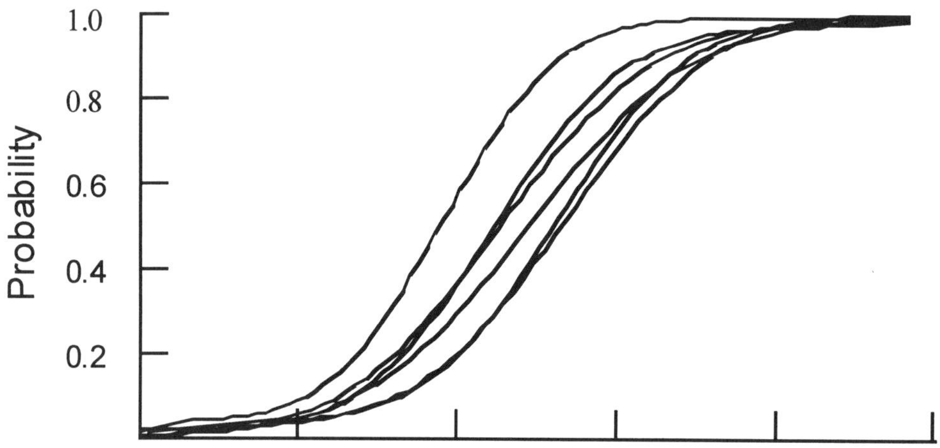

Fig. 4.8. A sampling design which nests sampling for Type A uncertainty (variability) within sampling for Type B uncertainty (lack of knowledge about constant but unknown parameters) yields a family of CDFs for the output distribution. The variability between the CDFs represents the uncertainty in the central tendency and shape of the distribution of the output due to lack of knowledge.

The choice of the number of simulations to be run in a Monte Carlo analysis depends on the goals for the analysis and on practical considerations, such as the time required to perform the analyses. Providing a good estimate of the mean of an output variable generally requires fewer simulations than does an equivalent estimate of its variance. If the mean value is of interest, then an approximation for the number of simulations can be derived given an estimate of the standard deviation and a desired value for the standard error. The problem of estimating sample size is more complicated if one is interested in the tails of a distribution. The tails of a distribution often represent low frequency events, and thus a large sample size is usually needed to ensure good representation in the tails. For example, suppose that one is interested in the upper 5% of a distribution. Using simple random sampling the number of simulations, n, required to observe 1 value in the upper 5% of the distribution with 95% confidence is

$$0.95^n = 0.05 \tag{4.53}$$

$$n = \frac{\ln(0.05)}{\ln(0.95)} \cong 59 \tag{4.54}$$

This can be interpreted as saying that if one wants to determine the position of the 95th percentile, with a 5% chance of the value being wrong, then one could perform 59 simulations, order the results, and use the largest value as the estimate of the percentile. Although the largest value is expected to lie within the upper 5% of the distribution it could occur anywhere within that quantile, i.e. it might lie within the upper 0.01% quantile. To reduce the chance to 1% of underestimating the 95th percentile requires 86 simulations. A 95% confidence interval can be constructed for the 95th percentile of any distribution using the data from a minimum of 100 simulations, although the width of the interval will be reduced by running more simulations.

Estimating the number of simulations to bring the estimated standard deviation of a distribution within a specified confidence interval is more problematic than doing so for the mean because the Central Limit theorem does not apply. In most situations estimates of the mean and variance of the output distribution will stabilize as the number of simulations increases, with the mean usually stabilizing sooner than the variance. One can set up a series of sets of simulations, say n simulations per set, and run enough sets to achieve a reasonably stable value of the variance. If the variance or mean fails to stabilize as n increases then the model structure and the parameters distributions should be critically examined to ensure that unreasonable parameters values are not being generated, such as values very near 0 that are used in denominators, or negative values for parameters which logically must be positive.

4.8 INTERPRETATION OF THE OUTPUT DISTRIBUTIONS

Descriptive statistics, described above for characterizing parameter distributions, are also used to characterize the uncertainty in the output variables. It is often quite useful to present plots of the cumulative distribution function and the probability density function, since the distributions of the output variables may be skewed or have two or more modes. Boxplots showing the median, interquartile range, and extreme percentiles are often informative (Warren-Hicks and Moore 1998). Although it is often possible to run thousands of Monte Carlo simulations, it is also true that the uncertainties in models results are often large, particularly when dealing with the projection of environmental impacts. Therefore, confidence intervals should accompany estimates of the mean or other parameters of the output distribution.

In risk assessment it is often not the mean of a distribution that is of interest but rather the location of various percentiles of the distribution. If one knows the parameters for a given distribution then the levels associated with any given percentile can usually be determined exactly. For example, the one-tailed 95th percentile for a standard normal distribution is 1.65σ from the mean, which means that 5% of the values in the population represented by the distribution would be greater that $\mu+1.64\sigma$. Usually, however, one will only have estimates of the parameters based on a random sample of the true distribution (Fig. 4.9). The uncertainty on those estimates depends on the size of the sample. Estimates of percentiles will have greater uncertainty than estimates of means. Tolerance intervals are used to represent regions that contain some proportion of the distributions with a specified level of confidence. In the risk assessment literature, such proportions of a distribution are sometimes mistakenly referred to as confidence intervals. However, a confidence interval expresses confidence in the estimate of a parameter of a distribution, whereas a tolerance interval identifies a region that contains a proportion of a distribution with some level of confidence (Hahn and Meeker 1991). The methods used to construct confidence intervals, such as the use of the t-distribution to compute the confidence interval for a mean, are not appropriate for constructing tolerance intervals.

There are two approaches typically used to estimate the percentiles for a distribution. In the first method one uses tabular values of the tolerance parameter, l, to estimate the percentiles. The tolerance parameter is used to compute the two-sided tolerance levels using the formula

$$t_p = X \pm l_{p,P} s \qquad (4.55)$$

where t_p is the upper p tolerance limit, X is the estimate of the mean, $l_{p,P}$ is the tolerance parameter for the two-sided percentile p and a confidence level of P, and s is the estimate of the standard deviation. One such table is available for normal distributions (Table 4.5). There are two probabilities to be dealt with when using the table. The first represents the percentile to be estimated and the second represents the confidence one wants to have in the estimate. Notice that this table reflects the slow convergence of the estimate of the variance with increasing sample size; even with 1000 degrees of freedom the tolerance statistic, l, does not converge with the theoretical value for infinite degrees of freedom.

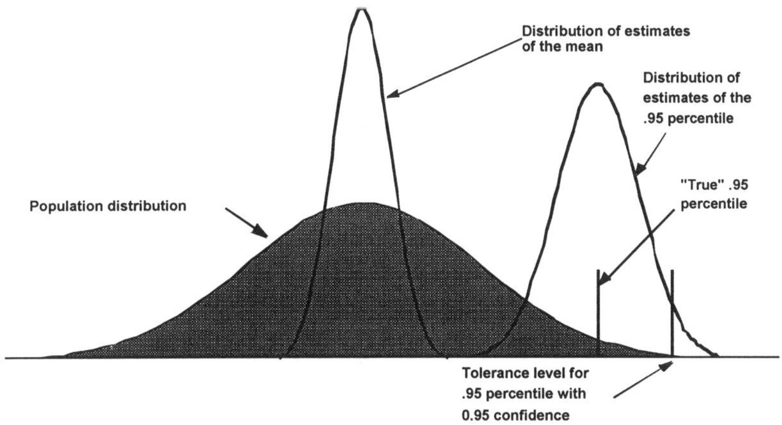

Fig. 4.9. When estimating percentiles for a distribution based on a sample, the uncertainties associated with the sample mean and variance contribute to the uncertainty in the location of the percentile.

For example, suppose one has estimated the mean and variance for a normal distribution based upon 21 samples and one is interested in estimating the level of the upper 0.95 percentile with a confidence of 0.99. The table is set up for two-sided tests. In the table one would select the right most group of columns because they represent the 0.99 confidence level, and then would select l = 2.62 from the 0.90 column for the proportion of the distribution. The 0.90 column is selected because only a one-sided level is desired. The level is then computed as

$$t_{0.95} = X + 2.62s \tag{4.56}$$

where $t_{.95}$ is the upper 0.95 tolerance limit, X is the estimate of the mean, and s is the estimate of the standard deviation.

Table 4.5. A Latin hypercube design for 5 parameters and 10 simulations. The numbers in the rows of the parameters show which quantile is sampled for the listed simulation. These values are assigned at random without replacement in each row.

Parameter	Simulation									
	1	2	3	4	5	6	7	8	9	10
A	4	7	1	10	8	2	5	9	3	6
B	1	8	2	4	6	3	10	5	7	9
C	7	3	6	9	2	5	8	1	4	10
D	5	4	3	6	9	1	2	7	10	8
E	9	6	2	5	1	4	7	8	10	3

The second method for constructing tolerance intervals, the distribution-free approach, can be used when one doesn't know if the assumption of normality is appropriate (Chapter 7). The distribution-free approach estimates the position of percentiles using order statistics. The order statistics are the data ordered from the smallest value, $x_{(1)}$, to the largest value, $x_{(n)}$. The distribution free point estimate of the pth percentile lies between $x_{(np)}$ and $x_{(np+1)}$ (Hahn and Meeker 1991). Therefore, the $(np)+1$ order statistic is used to estimate the location of the pth percentile, where (np) is the integer part of the product of n and p. For example, the point estimate of the 95th percentile for an n of 50 would be $x_{(48)}$. One tailed and two-tailed confidence intervals can be computed for percentiles if n is sufficiently large. The binomial distribution is used to compute the order statistics for a given confidence level and percentile. That method and tables of order statistics for estimating one- and two-tailed tolerance intervals can be found in Hahn and Meeker (1991).

The interpretation of a distribution of uncertainty depends on whether the distribution represents only variability (Type A uncertainty), or true uncertainty (Type B uncertainty), or an aggregation of both Type A and Type B uncertainty. The first case is equivalent to sampling a real population and is extremely unlikely to ever be encountered in practice. The mean, variance, median and other percentiles, skewness and kurtosis of the distribution would correspond to what would be expected from a sample of distribution of actual measurements. The second case is much more likely to be encountered in risk assessment. In this case, if there were no uncertainty, then the endpoint of the assessment would be represented by a single value rather than a distribution. For example, if one modeled the dose to an individual, then the distribution of uncertainty would represents the confidence that one has in the methods and information used to estimate the dose. The distribution would not represent variability that one would expect to see within a population, even if variability among members of a population had been used to estimate the uncertainty on some parameters.

Uncertainty analyses involving both Type A and Type B uncertainties are common in risk assessment. Aggregating variability and confidence, i.e. treating these sources of uncertainty as being equivalent produces a distribution of risk or dose that confounds systematic To illustrate this point in more detail consider an example in which the total dose, D, is equal to the sum of two components, A and B:

$$D = A + B \tag{4.57}$$

The first component, A, is a constant having a value of 120. The second component, B, varies among individuals and is represented by a normal distribution having a mean of 50 and a standard deviation of 15. The true distribution of the sum has a mean of 170 and a standard deviation of 15 (Fig. 4.10). In an assessment the unknown value of A is subjectively estimated to be between 70 and 130 with 100 being the most likely value. Assuming a triangular distribution, the true value of 120 falls just below the 95th percentile, and hence the distribution would be considered acceptable by many scientists. The distribution of estimated doses simulated using a Monte Carlo method shows the impact of the uncertainty in A (Fig. 4.10). The value associated with the 95th percentile of the simulated distribution is expected to result from a value of A and B near the upper tails of their respective distributions. In contrast, the median represents values for which one had only a moderate level of confidence in A and an average value for B, an A value of low confidence plus an infrequent, high value of B, etc. Thus one should have greater confidence that 95th percentile of the model result does not underestimate the true 95th percentile of the population than the confidence one has in stating that the median of the simulated distribution does not underestimate the median of the true distribution.

It should be noted that this example represents a worst-case but scientifically credible prediction because the most likely value of A was assumed to be lower than its true value. Had the distribution assigned to A ranged from 110 to 170 with a mode of 140, such that the true value was just within the lower bound of the 95% tolerance interval, then the distribution of simulated results would have been to the right of the true results.

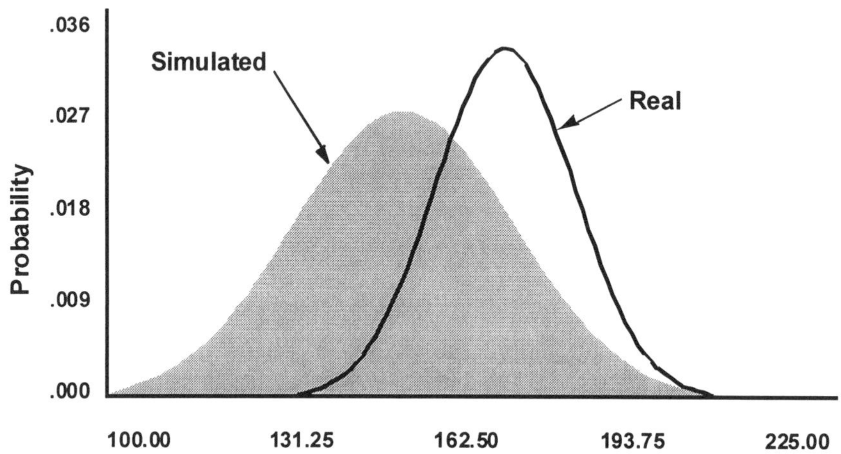

Fig.4.10. The distribution of simulated doses (shaded) lies to the left of the distribution of "real" doses.

4.9 SUMMARY

Uncertainty analysis shares many concepts and much terminology with classical statistics but it also deals with problems that typically cannot be addressed by classical statistics, such as the representation of systematic error due to lack of knowledge. Specification of subjectively derived distributions can be challenging, and methodologies for eliciting information from experts continue to improve. Analytical methods have largely been replaced by Monte Carlo methods for propagating uncertainty through models. Dynamic models still pose problems for some aspects of propagation of uncertainty, especially when parameters are stochastic through time. Although descriptive statistics generally work well for characterizing the distributions of the outputs from models, some important methods, such as the construction of tolerance intervals, are infrequently used in experimental statistics and hence unfamiliar to many scientists. The differences between classical statistics and uncertainty analysis can also lead to confusion when interpreting the results of uncertainty analyses, since the results are often presented using the same methods as one would use to describe distributions of measurements.

4.10 REFERENCES

American National Standards Institute.. U.S. Guide to the expression of uncertainty in measurement. ANSI/NCSL Z540-2-1997.National Conference of Standards Laboratories, Boulder, CO; 1997.

Bailar, J. C. III; Ederer, F. Significance factors for the ratio of a Poisson variable to its expectation. Biometrics 20:639-643; 1964.

Beach, L. R.; Braun, G. P. Laboratory studies of subjective probability: a status report. In Wright, G.; Ayton, P. Subjective probability. New York: Wiley & Sons; 1994.

Bogen, K. T. Methods to approximate joint uncertainty and variability in risk. Risk Assessment 15: 411-419; 1995.

Breshears, D. D.; Kirchner, T. B.; Otis, M. D.; Whicker, F. W. Uncertainty in predictions of fallout radionuclides in foods. Health Physics 57: 943-953; 1989.

Casman, E. A.; Morgan, M. G.; Dowlatabadi, H. Mixed levels of uncertainty in complex policy models. Risk Analysis 19: 33-42. 1999.

Clark, C. E. Importance sampling in Monte Carlo Analyses. Operations Research 9: 603-620; 1961.

Eberhardt, L. L.; Thomas, J. M. Designing environmental field studies. Ecological Monographs 61:. 53-73; 1991.

Edelmann, K. G.; Burmaster, D. E. Are all distributions of risk with the same 95th percentile equally acceptable? Human and Ecological Risk Assessment 2: 223-234; 1997.

Fishman, G. S. Monte Carlo concepts, algorithms, and applications. New York: Springer; 1996.

Gardner, R. H.; Cale, W.G.; O'Neill, R. V. Robust analysis of aggregation error. Ecology, 63: 1771-1779; 1982.

Graham, J. D.; Green, L. ; Roberts, M. J. In search of safety: Chemicals and cancer risks:. 80-114. Harvard University Press, Cambridge; 1988.

Hahn, G. J.; Meeker, W. Q. Statistical intervals: A guide for practitioners. New York: Wiley Interscience; 1991.

Hass, C. N. Importance of distributional form in characterizing inputs to Monte Carlo assessments. Risk Assessment 17: 107-113; 1997.

Hattis, D.; Burmaster, D. E. Assessment of variability and uncertainty for practical risk assessment. Risk Assessment 14: 713-730; 1994.

Helton, J. C. Treatment of uncertainty in performance assessment for complex systems. Risk Analysis 14:. 483-511; 1994.

Henrion, M.; Fischoff, B. Assessing uncertainty in physical constants. American Journal of Physics 54:. 791-798, 1986.

Hodak, D. G. Adjusting triangular distributions for judgemental bias. Risk Assessment 14: 1025-1031; 1994.

Hoffman, F. O.; Hammonds, J. S. Propagation of uncertainty in risk assessments: The need to distinguish between uncertainty due to lack of knowledge and uncertainty due to variability. Risk Analysis 14:. 707-712; 1994.

Hoffman, F. O.; Kaplan, S. Beyond the domain of direct observation: How to specify a probability distribution that represents the 'State of Knowledge' about uncertain inputs. Risk Analysis 19: 131-134; 1999.

Hora, S. C.; Iman, R. L. Expert opinion in risk analysis. The NUREG-1150 methodology. Nuc. Sci. and Eng. 102: 323-331; 1989.

Howard, R. A; Matheson, J. E. Readings in the principles and practices of decision analysis. Menlo Park, CA: Strategic Decision Systems; 1984.

International Atomic Energy Agency. Evaluating the reliability of predictions made using environmental transport models. Safety Series No. 100, Vienna, Austria: International Atomic Energy Agency; 1989.

Iman, R.L.; Conover, W. J. A distribution-free approach to inducing rank correlation among input variables. Communications on Statistics: Simulation and Computing 11: 311-334; 1982.

Kahnemann, D.; Slovic, P.; Tversky, A. eds. Judgement under uncertainty: Heuristics and biases, New York: Cambridge University Press; 1982.

Kirchner, T. B.; Whicker, F. W. Validation of PATHWAY, A simulation model of the transport of radionuclides through agroecosystems. Ecological Modeling 22: 21-44; 1984.

Kirchner, T. B. Estimation and application of uncertainty in assessment models. In Till, J. E.; Meyer, H. R. eds. Radiological assessment: A textbook for environmental dose analysis. In Press.

Law, A. M.; Kelton, W. D. Simulation modeling and analysis. New York: McGraw-Hill; 1982.

MacIntosh, D. L.; Suter II, G. W.; Hoffman, F. O. Uses of probabilistic exposure models in ecological risk assessments of contaminated sites. Risk Assessment 14: 405-419; 1994.

Mayo, D. G. Error and the growth of experimental knowledge. University of Chicago Press. 1996.

McKay, M. D.; Conover, W. J.; Beckman, R. J. A comparison of three methods for selecting values of input variables in the analysis of output from a computer code. Technometrics 21: 239-245; 1979.

Meyer, M.; Booker, J. Eliciting and analyzing expert judgement: A practical guide. New York: Academic Press; 1991.

Miller, D. R. Sensitivity analysis and validation of simulation models. Journal of Theoretical Biology 48: 345-360; 1974.

Morgan, M. G.; Henrion, M. Uncertainty: A guide to dealing with uncertainty in quantitative risk and policy Analysis. Cambridge: Cambridge University Press; 1992.

Mosleh, A.; Bier, V. On decomposition and aggregation error in estimation: Some basic principles and examples. Risk Analysis 12: 203-214;1992.

Mood, A. M.; Graybill, V; Boes, D. C. Introduction to the theory of statistics. New York. McGraw-Hill; 1974.

Naylor, T. H.; Balintfy, V; Burdick, D. S.; Chu, K. Computer simulation techniques. New York: John Wiley and Sons; 1968.

Naylor, T. H.; Finger, J.M. Verification of computer simulation models. Management Science, 14: 92-101; 1967.

National Council on Radiation Pprotection and Measurements.. A guide for uncertainty analysis in dose and risk assessments related to environmental contamination. NCRP Commentary No. 14, Washington, DC: National Council on Radiation Protection and Measurements; 1996.

O'Neill, R.V.; Rust, B. Aggregation error in ecological models. Ecol. Modelling, 7: 91-105; 1979.

Peterson, S. R.; Kirchner, T. B. Data quality and validation of radiological assessment models. Health Physics 74: 147-157; 1998.

Press,W. H.; Flannery,V.; Teukolsky, S. A.; Vetterling, W. T. Numerical recipes in C. Cambridge: Cambridge University Press. 735; 1988.

Sasser, L. B. The uptake and secretion of iodine-131 by lactating dairy cows resulting from a contaminated pasture. MS. Thesis, Colorado State University, Ft. Collins, CO; 1965.

Schlyakhter, A. I. An improved framework for uncertainty analysis: Accounting for unsuspected errors. Risk Analysis 14:. 441-447, 1994.
Shrader-Frechette, K. Mthodological rules for four classes of scientific uncertainty. In Lemons, J. Scientific uncertainty and environmental problem solving. Blackwell Science, Cambridge, Mass. 1996.
Seiler, F. A.; Alvarez, J. L. On the selection of distributions for stochastic variables. Risk Analysis 16: 5-18; 1996.
Smith, E. P.; K. Ye. A statistical perspective of uncertainty. In Warren-Hicks, W. J.; Moore, D. R. J. Uncertainty analysis in ecological risk assessment. SETAC Press. 1998.
Smith, A. E.; Ryan, P. B.; Evans, J. S. The effect of neglecting correlations when propagating uncertainty and estimating the population distribution of risk. Risk Analysis 12: 467-474; 1992.
Suter, G. W. Comments on the interpretation of distributions in "Overview of recent developments in ecological risk assessment". Risk Analysis 18: 3-4; 1998.
Tversky, A.; Kahneman, D. Judgement under uncertainty: Heuristics and biases. Science 185: 1124-1131; 1974.
Taylor, John R. An introduction to error analysis. Mill Valley, CA: University Science Books, Oxford University Press; 1982
USNRC. Severe Accident Risks: An assessment for five US nuclear power plants. Springfield, Virginia: U.S. Nuclear Regulatory Commission NUREG-1150 (National Technical Information Division); 1990.
Warren-Hicks, W. J.; Moore, D. R. J. Uncertainty analysis in ecological risk assessment. SETAC Press. 1998.
Wright, G.; Ayton, P. Subjective probability. New York: Wiley & Sons; 1994.
Zar, J. H. Biostatistical analysis. Englewood Cliffs, NJ: Prentice-Hall; 1984.

4.11 AUTHOR PROFILE

Dr. Thomas B. Kirchner is an experienced systems ecologist and modeler, having worked on a number of environmental problems including the effects of air pollution from coal-fired power plants and the transport of radionuclides through human food chains. He was involved with estimating the doses received by populations surrounding the Nevada Test Site due to tests of nuclear devices at that facility. Kirchner collaborated with Dr. F. Ward Whicker to construct the PATHWAY model to estimate ingestion doses and their uncertainties. The PATHWAY model has also been used to estimate the doses received by populations exposed to radionuclides from the Chernobyl accident. Kirchner has conducted extensive tests of validation, sensitivity analyses, and uncertainty analyses of the PATHWAY model. He was a participant in the Biospheric Model Validation Study, which was an international effort to test the efficacy of contaminant transport models. He also developed two commercially available applications for building and analyzing simulation models, PREMOD and Time-Zero, which were designed to enable students and non-programmers to build sophisticated ecological simulation models.

Chapter 5

APPLIED REGRESSION METHODS

Phillip L. Chapman

5.1 INTRODUCTION

5.1.1 Objectives

This chapter is a brief, non-mathematical introduction to regression methods. The presentation uses a series of examples to describe the methods and emphasizes the ideas of the procedures, rather than the computational formulas. Computations are performed using the statistical package MINITAB (MINITAB Inc., 1997), except the computations of the last example, which are done in SAS (SAS Institute Inc, 1989). The objective of this chapter to provide a basic understanding of simple linear and multiple linear regression analysis, along with instruction in MINITAB, so that the reader will be able to carry out regression analyses similar to those presented. The last example involves a more advanced nonlinear regression methodology that is not currently available in MINITAB. The objective in presenting that example to give the reader a head start when confronted with a nonlinear problem not suited to linear regression methods.

5.1.2 Terminology and notation

We begin by establishing some fundamental terminology. The term "regression methods" is applied to a group of widely used statistical techniques that relate a response of interest (Y) to one or more predictors (X's). The "regression" of Y on X is defined to be the average value of Y for a given value of X. In its simplest form the "regression" of Y on X is a line, as in Fig. 5.1. More generally, however, the regression of Y on X is a curve (Fig. 5.2). When there are two predictors, X_1 and X_2, the regression of Y on X_1 and X_2 is a surface that gives the average value of Y for each combination of values of X_1 and X_2. This surface may be a plane, as in Fig. 5.3, or a more complicated curved surface, as in Fig. 5.4. In general, the regression of Y on k predictors is a k-dimensional surface that gives the average value of Y for each combination of $X_1, X_2, \ldots, X_k$. The response Y is called the "dependent" variable, and the predictors (X's) are called the "independent" variables.

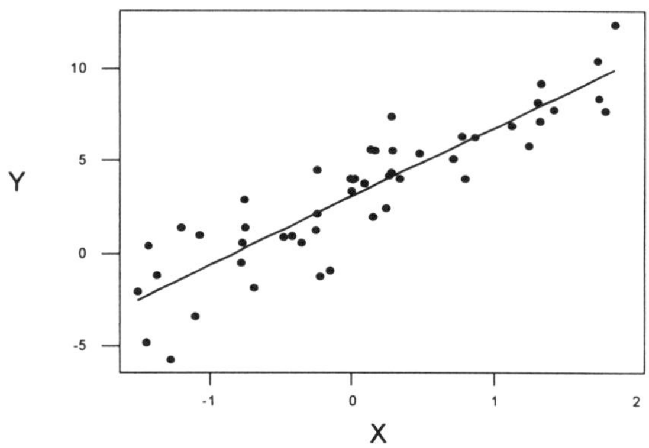

Fig. 5.1. Straight regression line: *Y* on *X*.

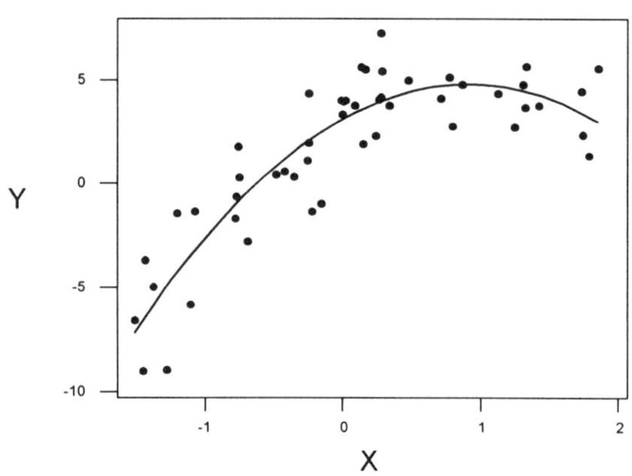

Fig. 5.2. Curved regression line: *Y* on *X*.

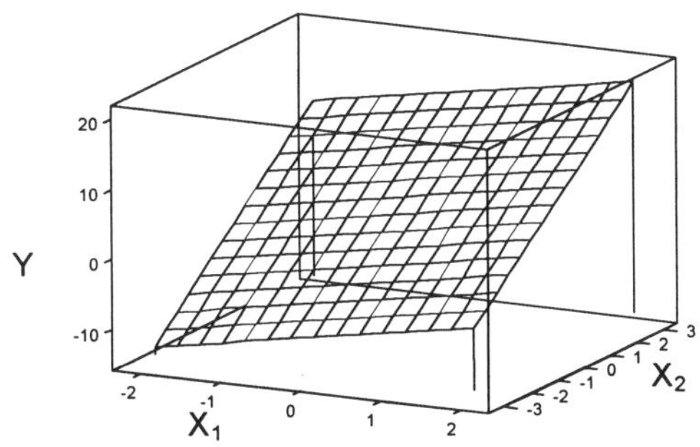

Fig. 5.3. Plane regression surface: Y on X_1 and X_2.

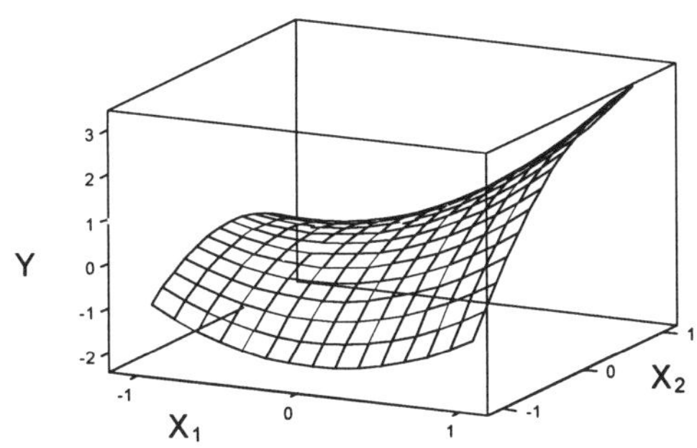

Fig. 5.4. Curved regression surface: Y on X_1 and X_2.

The success of a regression effort is judged relative to its objective. The objective may be *prediction* of the response based on values of the predictors. If so, success is judged by the closeness of the predictions to the observed responses, as measured by the squared prediction errors. If the objective is the discovery of the *form of the relationship* between the response and the predictors, the success of the effort is harder to judge, for the true form of that relationship will probably remain unknown. If the form of the relationship is known, the objective may be *estimation of the coefficients*, or parameters, of the model. The success of parameter estimation is also hard to judge, because calculation of the differences between the estimated and true parameters depends on knowing the true parameter values, and these are also likely to remain unknown.

A regression model is usually described as "simple regression" when there is a single predictor and "multiple regression" when there is more than one predictor. The term "multivariate regression" refers to studies in which two or more responses of interest are simultaneously modeled. Thus, a regression would be termed "univariate" when there is only one response variable, regardless of how many predictors are used. Because univariate regression is the most commonly used, the term "regression" (without qualifier) usually refers to that case. The use of regression terminology is not entirely consistent in the literature. For example, "univariate multiple regression" (one response, multiple predictors) is often included in books on multivariate analysis.

The term "linear regression" refers to models in which the relationship between the response and the predictors can be described as sum of terms, where each term is a parameter multiplied by a predictor. The most commonly used linear regression model is called "simple linear regression":

$$Y = \beta_0 + \beta_1 X + \varepsilon\,. \tag{5.1}$$

In this model the intercept parameter β_0 is multiplied by a constant predictor, 1, and the slope parameter β_1 is multiplied by the predictor X; then the two terms are summed. The ε term is called the "error", which is included to allow random deviations from the line. The usual assumptions are that the errors are independent and normally distributed with mean zero and constant variance (that does not depend on X). The following model would also be called a simple linear regression:

$$Y = \beta_0 + \beta_1 \sin(X) + \varepsilon\,. \tag{5.2}$$

In this model, sin(X) is the predictor, not X itself. The model involves a parameter multiple of sin(X), and is therefore linear. It is surprising to some that this model is called a linear regression, because a graph of Y versus X is certainly not a straight line.

The following model is a "multiple linear regression":

$$Y = \beta_0 + \beta_1 X_1 + \beta_2 X_2 + \varepsilon, \tag{5.3}$$

with predictors X_1 and X_2 . A slight modification of the above model gives a special case of multiple linear regression called "polynomial regression":

$$Y = \beta_0 + \beta_1 X + \beta_2 X^2 + \varepsilon, \tag{5.4}$$

with predictors X and X^2. Despite the fact that the graph of Y versus X is a parabola, this a *linear* regression because each predictor is multiplied by a parameter; then the terms are summed.

We follow the common statistical notation that uses upper case letters when referring to the random variables, Y, X_1, X_2, ..., X_k, and uses lower case letters with subscript i when referring to observed values of the ith data point, y_i, x_{i1}, x_{i2},..., x_{ik}. Lower case letters are also used to describe functional relationships, e.g. $y = ax + b$. Parameters are usually written using Greek letters (e.g.β_1), and estimates of parameters are written using the same Greek letter with a "hat" over it (e.g. $\hat{\beta}_1$).

5.1.3 Scope of the chapter

Now that the basic terminology is identified, we describe the scope of this chapter. The first example is a "simple linear regression." Using this example most of the fundamental ideas of parameter estimation, prediction and model checking are developed. The second example is "linearizeable"; that is, it is made linear by a transformation. The third example is a multiple regression with two predictors, which are squared or multiplied together to generate a total of five predictors. Using this example, issues relating to model selection, i.e. which predictors to retain in the model, are investigated. These first three examples will be analyzed using the statistical package MINITAB.

The last example is fundamentally nonlinear; that is, the prediction equation cannot be written as a sum of predictors, each multiplied by a parameter, and the prediction equation cannot be made linear by transformation. The computations of the nonlinear example are somewhat more complicated, because estimating the regression parameters uses an iterative procedure that requires input of initial guesses. However, from the user's point of view, the implementation is only slightly more complicated than linear regression. Because the calculations are not available in MINITAB, for this example we use SAS.

5.2 SIMPLE LINEAR REGRESSION

5.2.1 The "Subway" example

We introduce simple linear regression using an example that relates radon concentration (Bq m^{-3}) to relative humidity (%) in 108 air samples from an underground subway station in Seoul, South Korea, in 1990 (Kim, *et al.*, 1993). The data appear to indicate a weak positive relationship (Fig. 5.5).

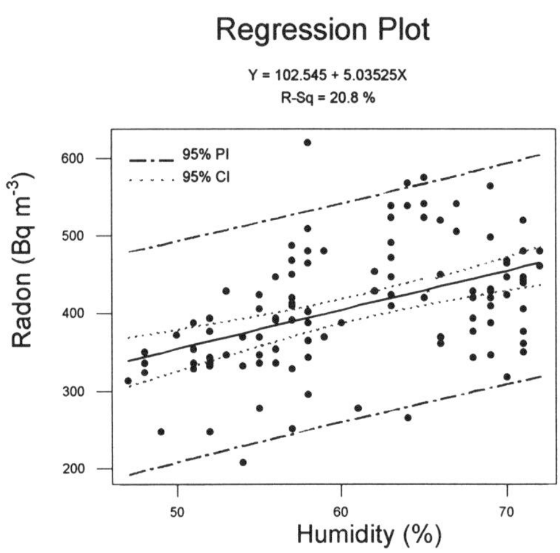

Fig. 5.5. Radon versus Humidity in 108 samples of the Subway data.

5.2.2 Least squares parameter estimation

The simple linear regression model assumes that the average value of radon (*Y*) for a given value of humidity (*X*) follows a straight line (Eqn. 5.1). By far the most common method of estimating the parameters of the line is the method of "least squares", which chooses parameter estimates that minimize the sum of squared vertical distances from the points to the line, the "Sum of Squared Errors" (*SSE)*:

$$SSE = \sum_{i=1}^{n} (y_i - (\beta_0 + \beta_1 x_i))^2 . \tag{5.5}$$

The values of β_0 and β_1 that minimize *SSE* are called the "least squares estimates" and denoted $\hat{\beta}_0$ and $\hat{\beta}_1$. The formulas for these estimates are obtained by setting the derivatives of *SSE* with respect to the parameters equal to zero, and then solving. The form of the equations is not of interest here because the calculations can be done by most statistical computing packages. The results, calculated by MINITAB, for these data are in Exhibit 5.1.

From the MINITAB output we obtain parameter estimates, $\hat{\beta}_0 = 102.54$ and $\hat{\beta}_1 = 5.0353$. The interpretation of $\hat{\beta}_1$ is that it is the estimated change in mean radon concentration *associated* with a one unit increase in relative humidity. We emphasize that the interpretation involves association between radon concentration and relative humidity, for there is no information in this analysis that establishes a causal relationship. The parameter estimate $\hat{\beta}_0$ is the estimated mean radon level when humidity equals zero. However, since the lowest humidity reading in the data set is 47%, the data contain no information about the form of the regression line for humidity values between 0% and 47%. Therefore, this interpretation of $\hat{\beta}_0$ requires an extrapolation of estimated properties of the regression line well beyond the range of the data. Such extrapolation is very risky, and should only be done in cases where there is a sound scientific argument to justify it. In the Subway example, the relationship between radon level and humidity is not understood

Exhibit 5.1. Basic MINITAB regression output for the Subway data

```
The regression equation is
Radon = 103 + 5.04 Humid

Predictor        Coef        StDev          T        P
Constant       102.54        58.59       1.75    0.083
Humid          5.0353       0.9542       5.28    0.000

S = 70.86        R-Sq = 20.8%       R-Sq(adj) = 20.1%

Analysis of Variance
Source            DF        SS         MS          F          P
Regression         1    139800     139800      27.85      0.000
Residual Error   106    532180       5021
Total            107    671980

Unusual Observations
Obs  Humid      Radon       Fit  St Dev Fit    Residual   St. resid
 19   54.0     207.00    374.45        9.53     -167.45      -2.38R
 37   64.0     266.00    424.80        7.40     -158.80      -2.25R
 83   58.0     621.00    394.59        7.39      226.41       3.21R
 96   65.0     577.00    429.84        7.82      147.16       2.09R
 98   64.0     569.00    424.80        7.40      144.20       2.05R
R denotes an observation with a large standardized residual
Predicted Values
Fit     StDev Fit          95.0% CI               95.0% PI
429.84       7.82    (414.33,   445.34)     (288.50,    571.17)
```

sufficiently to justify extrapolation. An exception to the prohibition against extrapolation would be lines that describe rate of radioactive decay, for which scientific arguments establish that the regression line should remain straight beyond the range of the collected data.

An estimate of the regression line is based on the estimated slope and intercept in the usual way:

$$\hat{y} = \hat{\beta}_0 + \hat{\beta}_1 x . \tag{5.6}$$

For the ith data point, $\hat{y}_i = \hat{\beta}_0 + \hat{\beta}_1 x_i$ is called the "predicted" value, because it is the *prediction* of Y_i at $x = x_i$. The term "fitted" is sometimes used in place of "predicted" to emphasize $\hat{y}_i$ as the height of the line that is being *fitted* to the observed data. Many authors use the terms "predicted" and "fitted" interchangeably.

The third model parameter is σ^2, the error variance, which is assumed to be the same ("homogeneous") for all values of *X*. Error variance is estimated by computing an "average" of the squared deviations from the estimated regression line; that is, by dividing the *SSE* by *n* - 2. The result is called "Mean Squared Error" (*MSE*), or $\hat{\sigma}^2$:

$$\hat{\sigma}^2 = MSE = SSE/(n-2) = 532180/(108-2) = 5021. \quad (5.7)$$

The divisor is *n* - 2, rather than *n*, because the squared deviations from the *estimated* regression line tend to be slightly smaller than the deviations from the *true* regression line. (This is the same reason the divisor in the computation of sample variance is *n* - 1, rather than *n*.)

MINITAB gives *n* - 2, *SSE*, and *MSE* on the "Residual Error" line of the "Analysis of Variance" portion of the output in Exhibit 5.1. The abbreviations used are DF (degrees of freedom), SS (sum of squares), and MS (mean square). The estimated error standard deviation is $\hat{\sigma} = \sqrt{\hat{\sigma}^2} = \sqrt{5021} = 71$. This value can be interpreted as an estimate of the standard deviation of radon measurements at a fixed value of relative humidity.

A statistic that is used to describe the overall quality of the fit of the regression line is the "coefficient of determination", R^2. In simple linear regression, R^2 is just the square of the usual Pearson product moment correlation, *r*. A more general, and more interpretable, definition of R^2 can be based on a breakdown of the total variability in *Y* into "explained" and "unexplained" portions:

$$\sum_{i=1}^{n}(y_i - \bar{y})^2 = \sum_{i=1}^{n}(\hat{y}_i - \bar{y})^2 + \sum_{i=1}^{n}(y_i - \hat{y}_i)^2. \quad (5.8)$$

The summation on the left is the "Total Sum of Squares" (*SST*), and can be recognized as the sum of squared deviations of observed y_i values about their mean. *SST* describes the *total* variability in the y_i values. The summation on the far right is the *SSE*, with $\hat{y}_i$

written in place of $\hat{\beta}_0 + \hat{\beta}_1 x_i$. *SSE* is the variability *not explained* by the regression model. The summation in the middle is the "Regression Sum of Squares" (*SSR*), which is the sum of squared deviations of the predicted values about their mean. (The mean of the predicted values equals $\bar{y}$.) *SSR* is the variability *explained* by the regression model. With these terms, Eqn. 5.8 can be rewritten:

$$SST = SSR + SSE \ . \tag{5.9}$$

R^2 can then be defined as the proportion of variability explained, or one minus the proportion of variability not explained:

$$R^2 = \frac{SSR}{SST} \quad \text{or} \quad R^2 = 1 - \frac{SSE}{SST} \ . \tag{5.10}$$

Taking the sums of squares from the "SS" column in the MINITAB output, we calculate for the Subway data:

$$R^2 = \frac{139800}{671980} = 0.208 \quad \text{or} \quad R^2 = 1 - \frac{532180}{671980} = 0.208 \tag{5.11}$$

R^2 is often interpreted as a percentage: 20.8% of the variability in radon concentration is explained by the linear regression on humidity.

5.2.3 Hypothesis tests about regression parameters

5.2.3.1 Hypothesis tests about the intercept and slope. If the regression errors (ε's) are independent and normally distributed, hypothesis tests about either β_0 or β_1 can be based on a *T*-statistic, which is formed in general as follows:

$$T = \frac{\text{estimate - hypotheiszed value}}{\text{std.error(estimate)}} . \tag{5.12}$$

The null hypothesis (H_0) that the slope is zero is usually the primary interest, because it is equivalent to the hypothesis of zero correlation, i.e. no linear relationship between the *X* and Y variables. H_0 is usually tested against the "two-sided" alternative that the slope is not zero (H_1), i.e.:

$$H_0: \beta_1 = 0 \text{ versus } H_1: \beta_1 \neq 0. \quad (5.13)$$

The T-statistic for the hypothesis test (5.13) is given in the MINITAB output:

$$T = \frac{\hat{\beta}_1 - 0}{\text{std.error}(\hat{\beta}_1)} = \frac{5.0353 - 0}{0.9542} = 5.28 \ (P = 0.000). \quad (5.14)$$

The estimate, $\hat{\beta}_1$, is labeled "Coef", and the standard error of the estimate, $\text{std.error}(\hat{\beta}_1)$, is labeled "StDev". If the null hypothesis is true, the T-statistic is distributed as a Student's t distribution with n - 2 degrees of freedom (d.f). Therefore, H_0 is rejected if the absolute value of the T-statistic exceeds $t_{\alpha/2}$ (the area under the Student's t density to the right of $t_{\alpha/2}$). In the Subway example T = 5.28 exceeds $t_{0.05/2} = 1.98$ ($\alpha = 0.05$); therefore, H_0 is rejected in favor of H_1. Hypothesis tests about the intercept can be constructed similarly; however, in this example tests about the intercept are not of interest, because the intercept is well beyond the range of the data.

More useful than a hypothesis test *with any fixed* α is the "P-value" (or significance level), defined for a T-test as: the area under the Student's t density function, outside the interval $-T$ to T. The P-value can be reported as a summary of the evidence against H_0, in favor of H_1, without fixing α at any specific value, such as 0.05 or 0.10. The null hypothesis would be rejected for any value of α greater than the P-value, so the individuals using the reported results have complete information about the values of α for which H_0 would be rejected. In our example, the P-value in the output has been rounded to 0.000, but can be reported as "$P < 0.0005$". Therefore H_0 would be rejected for any $\alpha >$ 0.0005, which indicates very strong evidence that the slope is not zero.

MINITAB prints P-values for hypothesis tests with two-sided alternatives. Sometimes a hypothesis test with a one-sided alternative is preferred. For example, it may be well known that the relationship between Y and X cannot be negative, so only positive alternatives are considered:

$$H_0: \beta_1 \leq 0 \text{ versus } H_2: \beta_1 > 0. \quad (5.15)$$

H_0 is rejected if T exceeds t_α. In the subway example $T = 5.28$ exceeds $t_{0.05} = 1.65$, using the usual $\alpha = 0.05$; therefore, H_0 is rejected in favor of H_2.

The alternative hypothesis is sometimes called the "research hypothesis", because it represents the hypothesis that the researcher believes and wants to establish. Rejection of H_0 in favor of H_2 then "proves" (with probability of error controlled at α) the research hypothesis that the slope is greater than zero. Conversion of the two-sided P-value from MINITAB to a one-sided P-value uses the following rule: (1) If the estimate is in the direction of the alternative hypothesis, then divide the P-value by 2, (2) If the estimate is in the opposite direction of the alternative hypothesis, divide the P-value by 2, then subtract the result from 1.0. In our example, $\hat{\beta}_1 > 0$ is in the direction of the alternative hypothesis (H_2: $\beta_1 > 0$), so divide 0.0005, by 2, to obtain $P = 0.00025$. In this example the two-sided P-value and one-sided P-values are not meaningfully different. However, in many other examples, the one-sided P-value is less than a fixed α (i.e. reject), but the two-sided P-value is not less than the fixed α (i.e. do not reject). In those examples rejection of H_0 would depend on whether a one-sided or two-sided alternative is chosen.

5.2.3.2 Hypothesis tests about the mean of Y at $X = x_0$. In addition to hypothesis tests about the β's individually, hypothesis test can be constructed for any linear combination of the β's. The linear combination that is usually of interest is the mean of Y given X equals a specific value, x_0. That mean is $\beta_0 + \beta_1 x_0$, the height of the regression line, which is estimated by $\hat{y} = \hat{\beta}_0 + \hat{\beta}_1 x_0$. The test of the hypothesis that $\beta_0 + \beta_1 x_0$ equals some specific value, η_0, is based on a T-statistic that follows the general form given in Eqn. 5.12:

$$T = \frac{\hat{y} - \eta_0}{\text{std. error}(\hat{y})}. \tag{5.16}$$

One-sided and two-sided tests and P-values can be obtained in the same manner as any other T-test.

Optional MINITAB output will give the estimated height of line, $\hat{y}$ (the "fitted" or "predicted" value), for any specified x_0 value and the standard error of the fitted value, from which the T-statistic can be calculated. In the Subway example, consider the null hypothesis that mean Radon equals 400 ($\eta_0 = 400$) when Humidity equals 65 ($x_0 = 65$), versus a two-sided alternative hypothesis. At the bottom of Exhibit 5.1 the fitted value for $x_0 = 65$ is given as 429.84 with "StDev Fit" = 7.82, from which we calculate T:

$$T = \frac{429.84 - 400}{7.82} = 3.82 \tag{5.17}$$

Because T exceeds, in absolute value, $t_{0.05/2} = 1.98$, we reject (at $\alpha = 0.05$) the null hypothesis in favor of a two-sided alternative. The P-value can be computed using MINITAB. It is the area under the Student's t density function, for values of T greater than 3.82 and less than –3.82: $P = 0.0002$. We conclude that there is very strong evidence that mean radon, given humidity equals 65, is not equal to 400.

5.2.3.3 Hypothesis tests about the error variance. Hypothesis tests about the error variance (σ^2) can be constructed based on *SSE*. The null hypothesis that $\sigma^2 = 0$ is never of interest, because *any* deviation of the data points from a straight line would contradict that hypothesis. More often the interest is in a particular value, σ_0^2, versus a one-sided ("right") alternative:

$$H_0\text{: } \sigma^2 = \sigma_0^2 \quad \text{versus } H_1\text{: } \sigma^2 > \sigma_0^2 . \tag{5.18}$$

The test is based on a Chi-square statistic:

$$\chi^2 = \frac{SSE}{\sigma_0^2} . \tag{5.19}$$

If the null hypothesis is true, χ^2 is distributed as a Chi-square distribution, with degrees of freedom (d.f.) equal to n - 2. The null hypothesis is rejected if χ^2 exceeds χ_α^2, the upper 1-α point of the Chi-square distribution. (For a left alternative, H_2: $\sigma^2 < \sigma_0^2$, the null hypothesis would be rejected if χ^2 is less than $\chi_{1-\alpha}^2$.) In the Subway data, if $\sigma_0^2 = 3000$, then:

$$\chi^2 = \frac{532180}{3000} = 177.4 . \qquad (5.20)$$

Reject H_0 in favor of H_1, because 177.4 exceeds $\chi^2_{0.05}$ = 131.0. A much stronger statement is to note that the P-value (the area under the Chi-square density function beyond 177.4) is calculated by MINITAB to be 0.0000, which has been rounded and can be reported as $P < 0.00005$.

5.2.4 Confidence intervals for parameters

5.2.4.1 Confidence intervals for the intercept and slope. Confidence intervals for either β_0 or β_1 can be based on the Student's t distribution, and follow the general form:

$$\text{estimate} \pm \text{(value from t-table)} \times \text{std. error(estimate)} \qquad (5.21)$$

For the slope parameter, β_1, the $(1 - \alpha)\times 100\%$ confidence interval is:

$$\hat{\beta}_1 \pm t_{\alpha/2} \ \text{std. error}(\hat{\beta}_1) \qquad (5.22)$$

In the Subway example, a 95% ($\alpha = 0.05$) confidence interval for β_1 is:

$$5.0353 \pm 1.98 \times 0.9542 \quad \text{i.e.} \quad 4.15 < \beta_1 < 6.92 \qquad (5.23)$$

The interpretation of this interval is that it is calculated by a method such that under repetitions of the entire experiment, 95% of such intervals would include the true unknown slope. This interpretation is so awkward that investigators are tempted to modify it and say that there is a 0.95 probability that the true slope lies in this particular interval. The first interpretation emphasizes the parameter as a fixed unknown value that is included in the random interval at some controlled rate. The modified interpretation takes the interval as fixed, and makes a probability statement about the parameter, as if it were random. Strictly speaking, only the first interpretation is correct, but the modified interpretation is acceptable to some statisticians.

Regression calculations are always done as if the X's were fixed constants, rather than random variables. In many analyses, the X values are fixed by the experimenter, e.g. a radioactive decay model, in which the X is the time at which the experimenter takes data.

In the Subway data, X (humidity) is clearly a random variable, but the statistical analysis is performed "conditionally" on the values of X observed, that is, as if the X's had been fixed at those values. Standard errors formulas for $\hat{\beta}_0$ and $\hat{\beta}_1$ involve the sample standard deviation of the observed X's, and are interpreted "conditionally" on those X's values. Therefore, standard errors for $\hat{\beta}_0$ and $\hat{\beta}_1$ quoted will only be valid for the particular values of X observed. However, if a confidence interval procedure has correct coverage rate conditionally on X, then the same coverage rate holds unconditionally. A similar statement can be made for hypothesis tests. If a T-test has type I error rate α for each possible configuration of observed X values, it will have type I error rate α unconditionally.

5.2.4.2 Confidence intervals for the mean of Y at $X = x_0$. A confidence interval for the height of the regression line at $X = x_0$ can be computed using the general form for T-based confidence intervals given in Eqn. 5.21. The $(1 - \alpha)\times 100\%$ confidence interval for the height of the line at x_0 is:

$$\hat{\beta}_0 + \hat{\beta}_1 x_0 \pm t_{\alpha/2} \times \text{std. error}(\hat{\beta}_0 + \hat{\beta}_1 x_0) \, , \tag{5.24}$$

also written as:

$$\hat{y} \pm t_{\alpha/2} \times \text{std. error}(\hat{y}) \, , \tag{5.25}$$

where the degrees of freedom are n - 2. The formula for standard error is the same one that was used when computing hypothesis tests for the height of the regression line. For the Subway data, a 95% confidence interval for the height of the regression line when x = 65 is given in Exhibit 5.1 as 414.33 to 445.34.

A confidence interval is often computed for each value of x, and the results presented as a pair of confidence bands, as in Fig. 5.5. The shape of the bands can be understood by inspecting the formula for $\text{std. error}(\hat{y})$:

$$\text{std. error}(\hat{y}) = \sqrt{\frac{\hat{\sigma}^2}{n} + \frac{\hat{\sigma}^2 (x_0 - \bar{x})^2}{(n-1)s_x^2}} \, . \tag{5.26}$$

The first term (under the square root sign) can be recognized as the square of the "standard error of the mean"; it describes the variability in the estimation of the overall mean of Y. The second term is a scaled distance from x_0 to the center of the X's; it describes the variability due to estimation of the slope. Note that the variability due to estimation of the slope is greater for x_0 values farther from the mean of X, accounting for the hyperbolic appearance of the confidence bands.

The interpretation of confidence bands is that under repetitions of the experiment, intervals for the height of the line *at each* x, would include the true height of the line *at that* x 95% of the time. Note that the coverage rate for the intervals is claimed for only one x at a time. It would be incorrect to claim that 95% of the time the entire line would lie within the confidence bands. Methods have been developed for which such coverage could be claimed, but those methods produce bands that are wider than the ones presented here.

5.2.4.3 Prediction intervals for a future Y observation at $X = x_0$**.** The prediction of a future Y observation, Y_{new}, at $X = x_0$ is simply the estimated mean at that x_0, i.e. the height of the estimated regression line. Thus, the estimate of the mean and the prediction and of a new observation are the *same number*.

A prediction interval for Y_{new} at $x = x_0$ is similar to a confidence interval for mean of Y at $x = x_0$, because they are both centered at the same number, and because they both have similar interpretations in terms of hypothetical replications of the experiment. A 95% prediction interval is an interval calculated by a method such that under repetition of the experiment, and observation of Y_{new} values at $x = x_0$, 95% of such intervals would include the Y_{new} values. The interval is formed like the confidence interval for the height of the line, but with a slightly altered standard error formula:

$$\hat{y} \pm t_{\alpha/2} \times \text{std. error}(Y_{\text{new}} - \hat{y}) \tag{5.27}$$

$$\text{std. error}(Y_{\text{new}} - \hat{y}) = \sqrt{\hat{\sigma}^2 + \frac{\hat{\sigma}^2}{n} + \frac{\hat{\sigma}^2 (x_0 - \bar{x})^2}{(n-1)s_x^2}} \tag{5.28}$$

The standard error in the prediction interval formula is always larger than the standard error in the confidence interval formula (Eqn. 5.26), due to the additional $\hat{\sigma}^2$ term under the square root sign. This term is added to reflect the additional variability of Y_{new} from its mean. For the Subway data, a 95% prediction interval for a new Y at $x = 65$ is given in Exhibit 5.1 as 288.50 to 571.17. Prediction intervals can be computed for a new Y at every x on the entire line, and then displayed as prediction bands that look very much like the confidence bands of the previous section, but wider (Fig. 5.5.)

5.2.4.4 Confidence intervals for the error variance. A $(1 - \alpha)\times 100\%$ confidence interval for the error variance, σ^2, can be based on the Chi-square distribution, as follows:

$$\frac{SSE}{\chi^2_{\alpha/2}} < \sigma^2 < \frac{SSE}{\chi^2_{1-\alpha/2}} \quad \text{(d.f.} = \text{n} - 2) \tag{5.29}$$

A 95% confidence interval ($\alpha = 0.05$) for σ^2 in the Subway example is:

$$\frac{532180}{136.4} < \sigma^2 < \frac{532180}{79.4} \quad \text{i.e.} \quad 3902 < \sigma^2 < 6703 \tag{5.30}$$

A confidence interval for σ is obtained by taking the square root of the endpoints.

5.2.4.5 Relationship between hypothesis tests and confidence intervals. The confidence intervals described for β_1, β_2, the height of the regression line and σ, all have an additional useful property; that is, the confidence intervals contain exactly the parameter values that would *not* be rejected in a size α hypothesis test. Therefore, if an investigator reports result of an experiment in the form of a confidence interval, the reader of the report has a complete specification of null hypothesis parameter values that would, and would not, be rejected.

This property of confidence intervals is particularly helpful in identifying whether a completed experiment involved a sample size that is sufficient to meet the research objectives. For example, say the investigator believes the true slope to be about five, and wanted to demonstrate in a hypothesis test that the slope is non-zero, but the hypothesis test failed to achieve significance. That is, H_0: $\beta_1 = 0$ has not been rejected. If the confidence interval for the slope were –2 to 12, the investigator would conclude that the true slope may indeed be five, but the sample size was too low to exclude zero as a reasonable value. If the confidence interval for the slope were –2 to 2, the investigator would conclude that the sample size was sufficient to have excluded zero *if the true value had been five*. In the latter case, the sample size was adequate, but there just was no evidence that the true slope is different from zero.

Many investigators approach the above problem with what is called a "*pos hoc* power analysis", in which an attempt is made to estimate power (the probability of rejecting H_0, given a specific alternative), for various alternatives of interest. However, the logical argument and the computations are simpler when, as above, they are based on confidence intervals. Power analyses are usually not covered in introductory texts, but confidence intervals are.

5.2.5 Effects of failure of model assumptions

The validity of parameter estimates, hypothesis tests and confidence intervals depends on whether the assumptions of the simple linear regression models are satisfied. The simple linear regression model assumptions are summarized below:

A1: $E(\varepsilon_i) = 0$ (linearity)
A2: ε_i are mutually independent
A3: $Var(\varepsilon_i) = \sigma^2$ (variance same for all i: "homogeneity of variance")
A4: ε_i are normally distributed.

The failure each assumption has an impact, some more important than others. For example, it is critical that assumption A1 be at least approximately satisfied, because errors with non-zero expectation imply that the line $\beta_0 + \beta_1 x$ does not represent the mean of Y given x. That is, if the mean of ε is not zero, then $\beta_0 + \beta_1 x$ is not the regression line.

The assumption of independence (A2) is not critical to the *validity* of the least squares estimates of β_0 and β_1, but is critical to the *claimed accuracy* of those estimates. For example, we now note that the n = 108 readings in the Subway example were actually taken on only six dates. In Fig. 5.6 the six dates are coded with six different symbols. The appearance of the graph and the knowledge that neither radon nor humidity levels change rapidly within a day, suggest that readings taken on the same day are not independent of each other. However, it is reasonable to believe that readings taken weeks apart may be nearly independent. Therefore, data were averaged within each of the six dates. The results are plotted in Fig. 5.7, and the MINITAB regression results (n = 6) are given in Exhibit 5.2. The slope estimate from the analysis of averages is 5.161, nearly

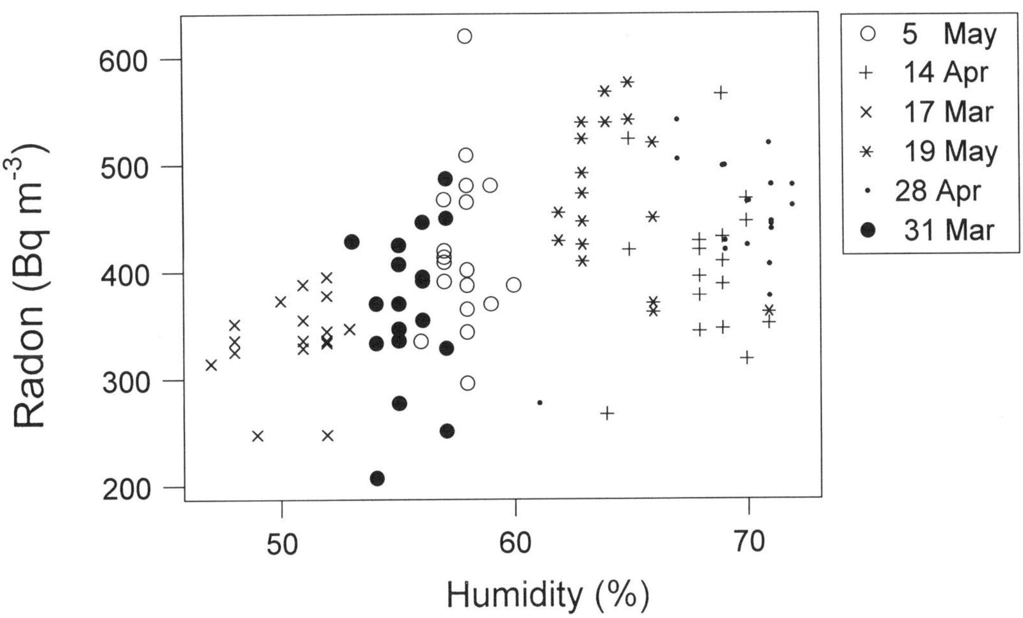

Fig. 5.6. Radon versus Humidity in the Subway example: points coded by sample collection date.

equal to the slope estimate from the original analysis (5.0353). But the standard error in the analysis of averages is more than twice as large, and the P-value is 0.063, not significant at $\alpha = 0.05$. The confidence interval for the slope, now computed using $n - 2 = 4$ degrees of freedom, is also much wider, -0.4576 to 10.78, and no longer excludes zero. In conclusion, acknowledging that the data contain only six independent pieces of information does not greatly change the *estimate* of the relationship between Radon and Humidity, but it greatly diminishes the *claimed accuracy of the estimate*, and the certainty with which it is claimed that the slope is non-zero.

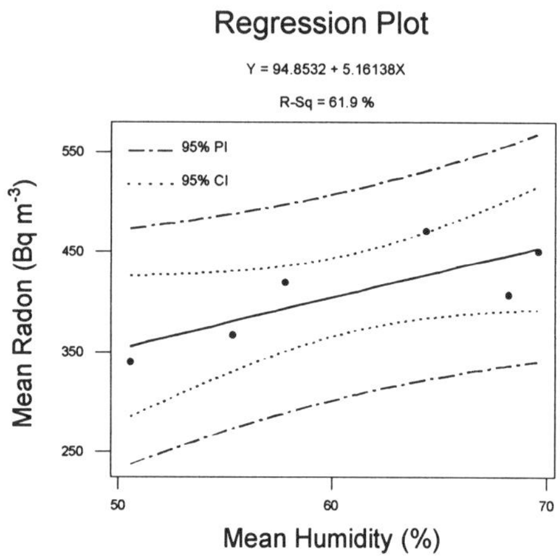

Fig. 5.7. Daily averages of Radon versus Humidity for the Subway data.

Exhibit 5.2. Basic MINITAB output for the radon and humidity averages by date (Subway data).

```
The regression equation is
y = 95 + 5.16 x

Predictor        Coef       StDev          T        P
Constant         94.9       124.2       0.76    0.488
x               5.161       2.024       2.55    0.063

S = 34.29       R-Sq = 61.9%     R-Sq(adj) = 52.4%

Analysis of Variance
Source          DF      SS      MS      F        P
Regression       1    7647    7647   6.50    0.063
Residual Error   4    4703    1176
Total            5   12350
```

The assumption of homogeneity of variance (A3) is also not critically important to the validity of the estimate of the slope. In fact, it can be shown that the least squares estimate of slope is unbiased, even if the variances are heterogeneous, rather than homogeneous. If the variances are heterogeneous, however, other less variable slope estimates are available based on a “weighted least squares” method that is discussed later, in the context of nonlinear regression. The standard errors, *T*-tests and confidence intervals for the slope will also be affected to some degree, but these procedures are “robust” in the sense that they are not greatly affected by the small to moderate failures of this assumption. The most important effect of heterogeneity of variance is on the validity of the prediction intervals. Since the regression method uses an estimate of error variance that is pooled from all of the data points, the estimate will represent some average value. Prediction intervals in regions where the variance has been underestimated would tend to include too few future observations. Similarly, prediction intervals in regions of small variability would tend to include too many future observations.

The assumption of normality (A4) of errors is also not critical for the purpose of making inference about the slope. The true distribution of the errors can deviate somewhat from normality, and the inference about the slope will be approximately accurate. However, distributions that are skewed, or heavy tailed, will have very inaccurate coverage rates for prediction intervals, particularly if one-sided intervals are used.

There are several tools that are commonly used to check for the failure of regression model assumptions. The most commonly used tools are plots of "residuals." The residuals are the differences between the observed values (y_i) and the fitted values ($\hat{y}_i$). Although residual plots are useful in simple linear regression, much of the information gained from residual plots can be obtained from the basic *Y* versus *X* plot. For example, a simple plot of *Y* versus *X* will usually indicate if the regression is curved, if the variance of the points from the line is non-constant (heterogeneous), or if there are unusual observations that should be checked. Residual plots are more important in multiple regression, and some examples of residual plots will be given in that section.

5.2.6 "Linearization" and the “Composition” example

A sample of unknown composition was exposed to a particle beam at an accelerator, and became activated. The only identifiable peak, observed with a gamma spectroscopy system, was at 0.511 MeV. This indicated that it was a positron emitter. In order to identify the radioisotope, the sample was counted at various times for ten-minute intervals. The data are shown in Table 5.1, along with a sample of background readings.

Table 5.1. The Composition data: Counts at 8 times, along with 10 background readings.

Bkg	Time	Counts
67	60	1546
74	70	1493
65	80	1544
64	90	1603
67	120	1437
58	150	1506
56	180	1412
73	360	1186
64		
65		
Ave=65.3		

An appropriate model for the count Y at time t would be:

$$Y = B + A_0 e^{-\lambda t} + \varepsilon , \tag{5.31}$$

where B is the background level, A_0 is the initial count rate, and λ is the decay rate, and ε is random error. This model is nonlinear, since the parameter λ and the predictor t are exponentiated. We now reformulate the model so that the parameters can be estimated using simple linear regression. Since B is estimated accurately relative to the other parameters (65.3 ± 1.78 s.e.m.), we take it as approximately constant, and subtract it from the response:

$$(Y - 65.3) = A_0 e^{-\lambda t} + \varepsilon . \tag{5.32}$$

This model can now be "linearized" by taking logarithms (base e):

$$\ln(Y - 65.3) = \ln(A_0) - \lambda t . \tag{5.33}$$

The decay rate λ can be estimated as the negative of the slope of the regression line, based on a simple linear regression of ln(Y - 65.3) versus t. A plot of ln(Y - 65.3) versus t is given in Fig. 5.8, and the MINITAB output from the linear regression is given in Exhibit 5.3. We note that if B cannot be assumed to be known, the model is not linearizable and must be treated as a nonlinear model. The nonlinear model methods of Section 5.4 must be used.

Exhibit 5.3. Basic MINITAB output for a "linearized" model of the Composition data.

```
The regression equation is
LnCount = 7.37 -0.000938 Time

Predictor        Coef       StDev         T        P
Constant      7.37075     0.02297    320.89    0.000
Time       -0.0009379   0.0001379     -6.80    0.000

S = 0.03593     R-Sq = 88.5%     R-Sq(adj) = 86.6%

Analysis of Variance

Source           DF        SS          MS          F        P
Regression        1   0.059724    0.059724     46.25    0.000
Residual Error    6   0.007748    0.001291
Total             7   0.067472

Unusual Observations
Obs   Time   LnCount    Fit   StDev Fit    Residual   St Resid
  8    360    7.0217 7.0331      0.0331     -0.0114    -0.81 X

X denotes an observation whose X value gives it large influence.
```

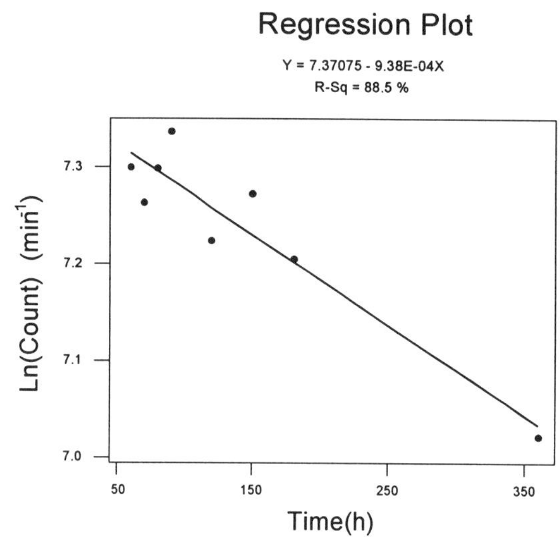

Fig. 5.8. Plot of ln(Count - 65.3) versus Time for the Composition data.

5.3 MULTIPLE LINEAR REGRESSION

5.3.1 The "Surface" example

Silicon drifted surface barrier detectors are efficient for measuring charged particles. One version of this has collecting electrodes on opposite edges of the detector. The total charge created by the incident particle is shared between the two collectors. If the detector is aligned properly, the two signals can be used to determine the location of the incident particle in a selected direction. A second detector placed behind the first detector and rotated 90° will give information in the vertical direction.

The division of voltage in a detector is not always linear, and the axes may not be exactly perpendicular. Thus, the system must be calibrated with a collimator that has a series of holes that are at precisely machined locations given by Y_1 (horizontal) and Y_2 (vertical) coordinates (marked with open circles in Fig. 5.9). When the detector is exposed to a beam of charged particles, only those particles passing through the holes will reach the detectors. The data consist of signals X_1 (horizontal) and X_2 (vertical), (solid circles in Fig. 5.9), that should represent the coordinates of the holes. The objective is to calibrate the system by finding an equation that transforms measured X values into correct Y values. We will consider second order polynomial functions of the form:

$$y = \beta_0 + \beta_1 x_1 + \beta_2 x_2 + \beta_3 x_1 x_2 + \beta_4 x_1^2 + \beta_5 x_2^2 + \varepsilon . \qquad (5.34)$$

A separate model will be fit to predict the horizontal (Y_1) and vertical (Y_2) components. Our model can be considered a multiple regression with $k = 5$ predictors. If both Y's are considered together, the problem would be called a "multivariate" multiple regression. However, there is little to be gained by modeling both Y's together, so we will stay with the "univariate" multiple regression.

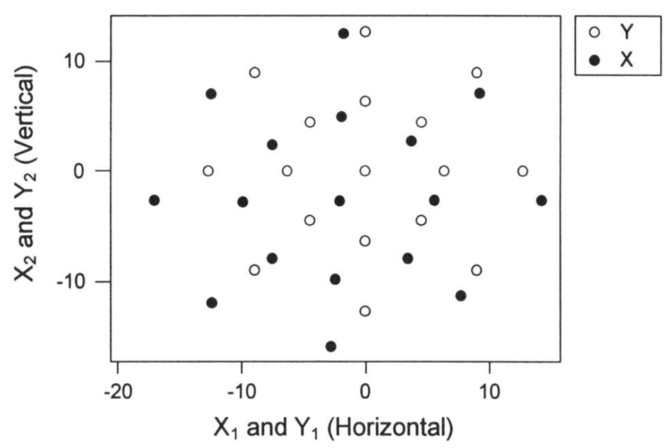

Fig. 5.9. Machined locations (Y_2 versus Y_1: open circles) and particle locations (X_2 versus X_1: solid circles) for the Surface data.

Our model is non-standard in one sense. In the usual regression analysis, the predictors are considered to be fixed values, measured without error, and the responses are random quantities that deviate from the fixed surface by the random error, ε. In the Surface problem, however, the Y's are fixed because they are precisely machined and precisely measured, and the X's represent the random response. That is, we have reversed the roles of predictor and response; the predictors are random and the responses are fixed. The "Classical" approach to calibration is fit a model that predicts the readings from the instrument being calibrated (random responses), based on the fixed predictors (known true values), and then to *invert* the system of model equations. Given a new value from the instrument being calibrated, the inverted equations are then used to estimate the unknown true value. We have taken the "Inverse Prediction" approach to calibration, which treats the known true values as the response and the instrument readings as the predictors, then estimates the prediction equations directly by regression. There is a large literature debating the merits of the two approaches, but the two approaches give nearly the same result when the fit of the regression equation is very good. It is only when the calibration relationship is weak that the two methods given noticeably different results.

5.3.2 Least squares parameter estimation, hypothesis tests about parameters, confidence intervals and prediction intervals in multiple regression

In this subsection we cover several topics at once by describing methods for multiple regression that are extensions of methods already presented for simple linear regression. The extensions are typically straightforward. For example, the general form of the multiple linear regression model is:

$$Y_i = \beta_0 + \beta_1 X_{i1} + \beta_2 X_{i2} + \ldots + \beta_k X_{ik} + \varepsilon_i . \tag{5.35}$$

The X's are now double subscripted; the first subscript, *i,* indicates which data point, and the second subscript, *j,* indicates which predictor. The β_j parameters in the multiple regression model are called the "partial regression coefficients". They give the average change in Y for a unit change in X_i, *all other X's held constant.* It is important to understand that this does not imply that changing X_j by one unit will *cause* a change in Y, because a change in X_j may be associated with changes in the other X's.

Least squares estimation is achieved by finding parameter estimates that minimize the sum of squared errors:

$$SSE = \sum_{i=1}^{n} (y_i - (\beta_0 + \beta_1 x_{i1} + \beta_2 x_{i2} + \ldots + \beta_k x_{ik}))^2 \tag{5.36}$$

Formulas for parameter estimates are found by differentiating *SSE* with respect to the β's, setting the system of equations equal to zero, and solving. Matrix equations for parameter estimates, standard errors of parameter estimates, fitted values, standard errors of fitted values, and standard errors for prediction errors all involve inverting a $k+1$ by $k+1$ matrix, but are easily handled by MINITAB. Predicted values ($\hat{y}_i$) are calculated by putting the parameter estimates into the model, Eqn. 5.35.

An overall assessment of the fit of the regression can be based on the "coefficient of determination," R^2. The calculation of R^2 follows the same formula, Eqn. 5.10, as in simple linear regression. The interpretation is also nearly the same: R^2 is the proportion of the variability in *Y* that is explained by the linear regression on the *X's*. Another interpretation of R^2 that applies to both simple linear and multiple linear regression is that R^2 is the square of the Pearson product moment correlation, *r*, between the observed data (y_i 's) and the predicted values ($\hat{y}_i$'s). A common question is: "How large should R^2 be?" Of course, there is no real answer to that question. It is much like the question: "How much money is enough?" Investigators working with physical systems are often not satisfied unless R^2 is nearly 1.0. Investigators working with biological systems are often satisfied with R^2 values in the range 0.3 to 0.5.

The construction of *T*-tests, confidence intervals and prediction intervals in multiple regression follows the same pattern as for simple linear regression. The only important difference is that the degrees of freedom are now $n - k - 1$, (number of data points minus number of β's). The formula for degrees of freedom from simple linear regression ($n - 2$) is easily seen to be a special case of the new formula, with $k = 1$. The basic MINITAB output for predicting Y_1, using the second order model of Eqn. 5.34 is given in Exhibit 5.4.

Exhibit 5.4. MINITAB output for the Surface example: predicting Y_1 using a second order polynomial model in X_1 and X_2.

```
The regression equation is
y1 = 1.70 + 0.815 x1 - 0.0297 x2 - 0.00295 x1x2 - 0.00219 x12 +
0.00085 x22

Predictor        Coef       StDev          T        P
Constant       1.7009      0.1190      14.30    0.000
x1           0.814896    0.008789      92.72    0.000
x2          -0.029712    0.009804      -3.03    0.011
x1x2        -0.002950    0.001309      -2.25    0.046
x12        -0.0021946   0.0008949      -2.45    0.032
x22          0.000845    0.001093       0.77    0.456

S = 0.2665      R-Sq = 99.9%      R-Sq(adj) = 99.9%

Analysis of Variance

Source          DF        SS          MS          F        P
Regression       5    805.65      161.13    2268.41    0.000
Residual Error  11      0.78        0.07
Total           16    806.43

Source        DF      Seq SS
x1             1      803.95
x2             1        0.70
x1x2           1        0.37
x12            1        0.58
x22            1        0.04

Unusual Observations
Obs    x1      y1        Fit   StDev Fit   Residual   St Resid
17    7.7  8.9800     8.5459      0.1839     0.4341      2.25R

R denotes an observation with a large standardized residual

Predicted Values

     Fit  StDev Fit         95.0% CI              95.0% PI
  0.1086     0.1198   ( -0.1551,   0.3722)( -0.5346,   0.7517)
 -8.3405     0.1845   ( -8.7467,  -7.9343)( -9.0540,  -7.6270)
```

Estimates for each β parameter are given, along with T-tests for the null hypothesis that each parameter is zero, versus a two-sided alternative. Confidence intervals for β parameters are not given, but can be formed using the given standard errors (labeled "StDev") and an appropriate value from the Student's t table. R^2 is given as a percent; 99.9% of the variability in Y_1 is explained by the second order polynomial model in X_1 and X_2.

Confidence intervals for the height of the regression surface are not part of the basic output, but have been requested for two X points that are not in the data set. The first point is (-2.02, -2.43), a point at the center of the X's. Results are at the bottom of Exhibit 5.4. The estimated mean of Y_1 at this point is 0.1086, with standard error equal 0.1198. The 95% confidence interval, -0.1151 to 0.3722, is given by MINITAB, calculated using the same formula used for simple linear regression (Eqn. 5.22), but with the appropriate standard error from the multiple regression substituted. A prediction interval for the same X point is also given, -0.5346 to 0.7517. As in the case of simple linear regression, the prediction interval is wider than the confidence interval at the same X point, because the prediction interval equation contains an extra $\hat{\sigma}^2$ to account for the variability of a Y_{new}. The second X point for which a confidence interval for the height of the regression surface is obtained is (-12.0, -12.0), a point near the lower left corner of Fig. 5.9. The estimated mean of Y_1 is at this point is –8.3405, with standard error 0.1845. For this point, the standard error is larger, and the confidence and prediction intervals are wider, than for the first X point near the center. Thus, estimates are more variable, and confidence and prediction intervals are wider near the edges of the X's than in the center of the X's, just like in simple linear regression.

5.3.3 Model selection methods

It can be demonstrated that parameter estimates and predictions are more variable when unneeded terms are included in a multiple regression model. For this reason, as well as the principle of simplicity (called "parsimony" in modeling circles), we attempt to use the simplest model that will satisfy the experimental objectives. The process of deciding which potential predictors to include and not to include in the model is called "model selection". Model selection is particularly difficult when the potential predictors are highly correlated, that is, when they provide overlapping information about the response. In that case, a particular predictor may appear important when included in the model with one set of predictors, yet may appear unimportant when included in the model with another set of predictors. The problem of highly correlated potential predictors is called the problem of "collinearity". In the Surface example, the predictors X_1 and X_2 can be seen in Fig. 5.9 to be nearly uncorrelated, so there is no important collinearity problem in our example. There is, however, some correlation between the X's and their squared and interaction terms.

Although the subject of model selection is too broad to cover here, we will describe briefly some of the most commonly used model selection methods and demonstrate two of them using the Surface data.

1. Forward selection – start with a very small model (often just the intercept), and at each step add the *most* significant (lowest P-value, or highest F) single potential predictor available, if it satisfies a minimum significance level.
2. Backward elimination – start with a very large model, and at each step remove the *least* significant predictor in the model, if it does not satisfy a minimum significance level.
3. Stepwise selection – a hybrid between forward selection and backward elimination. Use forward selection, but at each step remove the least significant term, if it does not satisfy a minimum significance level.
4. "Best" model based on some criterion. Three of the most commonly used criteria are given below:
   A. Minimum Mallows Cp - attempts to minimize total expected squared prediction error.
   B. Highest adjusted R^2 – selects the model that explains the largest proportion of the variability in Y (with an adjustment penalty for the number of predictors).
   C. Minimum Akaike Information Criterion (AIC) – a general information-theoretic criterion applicable to many nonlinear, as well as linear, models.

5.3.3.1 Model selection for Y_1 in the Surface data. We begin model selection for the Surface data by doing a backward elimination ($\alpha = 0.05$) starting with the second order model for Y_1 (Eqn. 5.34). Inspecting the parameter estimates in Exhibit 5.4, we see that the maximum P-value among the predictors is associated with X22 (x_2^2). After removing the X22 term, we refit the model. The results are given in Exhibit 5.5. All terms are now significant at $\alpha = 0.05$, so the process stops, and the model in Exhibit 5.5 is the final model. In this model R^2 is 99.9%. Very little appears to have been lost by removing x_2^2 from the model.

Exhibit 5.5. Final model results from Backward elimination model selection for the Surface data (x_2^2 removed from the model).

```
The regression equation is
y1 = 1.76 + 0.814 x1 - 0.0327 x2 - 0.00286 x1x2 - 0.00241 x12

Predictor        Coef        StDev          T        P
Constant      1.76121      0.08833      19.94    0.000
x1           0.814200     0.008595      94.73    0.000
x2          -0.032687     0.008865      -3.69    0.003
x1x2        -0.002858     0.001281      -2.23    0.046
x12        -0.0024144    0.0008342      -2.89    0.013

S = 0.2620        R-Sq = 99.9%      R-Sq(adj) = 99.9%
```

MINITAB has a Stepwise model selection option. The criterion for inclusion in the model is an *F*-test, which is equivalent to the *T*-test we have been using. The *F*-statistic is equal to T^2, and the null hypothesis that a parameter is zero is rejected if *F* exceeds a value in a table of the percentiles of the F distribution. The MINITAB user specifies the F-table value, which can be chosen to be approximately equivalent to a P-value criterion. The Stepwise model selection procedure can be converted to a Backward elimination procedure by setting the F-value required to add a variable at 100000, so that variables can only be removed from the model.

MINITAB also has a "Best subsets regression" option that identifies models with the highest R^2 for each number of predictor variables. It prints R^2, Mallows *Cp*, and adjusted R^2 for each model. The user can then select the "best" model based on any of those criteria. A "Best subsets regression" model selection for Y_1 of the Surface data is given in Exhibit 5.6. The model with the lowest *Cp* is the model previously selected by Backward elimination; it is the model that contains all terms except x_2^2.

A common question is: "Why not pick the model with the highest R^2?" Although this idea has intuitive appeal, it does not work, because models with more predictors typically have larger values of R^2. The reason can be seen by inspecting the formula for *SSE* in Eqn. 5.36. Model parameters are estimated by minimizing *SSE* over all possible parameter values. Fitting a model with predictors omitted is equivalent to restricting some $\hat{\beta}_i$ values to be zero, then minimizing *SSE* with respect to the others. The minimum *SSE* achieved by the *restricted* minimization cannot be lower than the *SSE* achieved by the *unrestricted* minimization. Therefore, R^2 for the model with fewer predictors cannot be higher than the R^2 for the model with more predictors. R^2 can only be used to compare models with the *same number of predictors*.

The other model selection criteria (adjusted - R^2, Mallows *Cp,* and Akaike's *AIC)* can be used to compare models with differing numbers of predictors. They all attempt to balance the improved fit resulting from adding another predictor with some kind of penalty for making the model more complicated. However, all of these criteria are based on *SSE;* therefore, when comparing models with the same number of predictors, all of these model selection criteria will pick the same "best" model. When comparing models that differ by a single predictor, the adjusted - R^2 criterion will be the most liberal about retaining that predictor in the model. Using adjusted - R^2 as a selection criterion, the jth predictor will be retained in the model if the *T*-statistic (for testing H_0: $\beta_j = 0$) exceeds 1.0. Similarly, using Mallows *Cp* as a selection criterion, the jth predictor will be retained in the model if the *T*-statistic exceeds $\sqrt{2}$.

Exhibit 5.6. MINITAB "Best subsets" regression results for Y_1 of the Subway data.

Response is y1

Vars	R-Sq	Adj. R-Sq	C-p	s	x1	x2	x1x2	x12	x22
1	99.7	99.7	21.8	0.40607	X				
1	13.4	7.6	9820.6	6.8240				X	
2	99.8	99.7	13.9	0.35577	X	X			
2	99.8	99.7	14.9	0.36258	X			X	
3	99.9	99.8	7.4	0.29939	X	X		X	
3	99.8	99.8	10.7	0.32804	X	X	X		
4	99.9	99.9	4.6	0.26202	X	X	X	X	
4	99.9	99.8	9.1	0.30852	X	X		X	X
5	99.9	99.9	6.0	0.26652	X	X	X	X	X

5.3.3.2 Model selection for Y_2 in the Surface Data. The results of a Backward elimination model selection for Y_2 (with "F to remove" = 4) are given in Exhibit 5.7. There are four steps in the procedure; the model including only x_2 and x_2^2 is selected.

Exhibit 5.7. Backward elimination selection results for Y_2 in the Surface data.

```
F-to-Enter:   10000.00     F-to-Remove:        4.00

 Response is      y2     on  5 predictors, with N =    17

    Step            1         2         3         4
Constant        2.458     2.443     2.455     2.452

x1             -0.007    -0.006
T-Value         -0.52     -0.53

x2              0.892     0.892     0.892     0.889
T-Value         60.34     63.30     65.31     67.77

x1x2           0.0012    0.0012    0.0016
T-Value          0.63      0.65      0.92

x12           -0.0002
T-Value         -0.12

x22           -0.0048   -0.0047   -0.0047   -0.0046
T-Value         -2.90     -3.15     -3.24     -3.20

S               0.402     0.385     0.374     0.372
R-Sq            99.78     99.78     99.77     99.76
```

The results of "Best subsets regression" for Y_2 are given in Exhibit 5.8. They indicate that the model with the lowest Mallows *Cp* involves only the predictors x_2 and x_2^2, yet achieves an R^2 of 99.8%. There is no evidence that any terms involving x_1 are needed in the model. The minimum Mallows *Cp* criterion and the Backward elimination method both selected the same model.

Exhibit 5.8. MINITAB "Best subsets" regression results for Y_2 of the Subway data.

Response is y2

Vars	R-Sq	Adj. R-Sq	C-p	s	x1	x2	x1x2	x12	x22
1	99.6	99.6	7.8	0.47268		X			
1	20.9	15.7	3937.7	6.5197					X
2	99.8	99.7	1.0	0.37204		X			X
2	99.6	99.5	8.7	0.47634		X		X	
3	99.8	99.7	2.3	0.37418		X	X		X
3	99.8	99.7	2.4	0.37624	X	X			X
4	99.8	99.7	4.0	0.38490	X	X	X		X
4	99.8	99.7	4.3	0.38941		X	X	X	X
5	99.8	99.7	6.0	0.40173	X	X	X	X	X

5.3.3.3 "Hierarchical" model selection. Selection by the lowest Mallows *Cp* criterion and Backwards elimination methods have been demonstrated using a second-order polynomial regression of the Surface data. However, when fitting polynomial models, many statisticians (including this one) choose to select only polynomial models that are "hierarchical" in the sense that each lower-order term would be included in the model whenever a higher-order term that contains it is included. Using this hierarchical restriction, it would not be possible to have a model with x_1^2 in the model, but not x_1. It would also not be possible to have x_1x_2 in the model, without having x_1. Stepwise model selection in MINITAB does not observe any of these hierarchical restrictions; however, if there are not too many predictors, hierarchical polynomial models can be selected by careful inspection of "Best subsets regression" output, and by manipulation of the MINITAB stepwise procedures. In the Backward elimination procedure for Y_2 of the Surface example (Exhibit 5.7), x_1^2 was removed first, then x_1, then x_1x_2. This model selection was not hierarchical at Step 2, because x_1 was removed while x_1x_2 remained in the model. A separate Backward elimination model selection (by hand) in which the hierarchical restrictions were observed was performed on the Surface data and produced the same final model chosen in Exhibit 5.7.

There is no general agreement on which is the best method of model selection. Many researchers try several methods, and find comfort when, as in the Surface example, more than one method chooses the same model. Models selected by automatic procedures should be taken as a starting point. Top candidate models from these procedures are then checked and compared before a final decision is made.

5.4.4 Checking model assumptions with the residual plots

The model assumptions in multiple regression are the same as in linear regression. The errors are assumed to be normal, independent, mean zero, and of homogeneous variance. However, in multiple regression evaluating the validity of model assumptions is harder than in simple linear regression. The single most useful tool is the plot of residuals versus predicted values. The residuals are defined as observed values minus fitted (or predicted) values:

$$r_i = y_i - \hat{y}_i . \tag{5.37}$$

The residuals are intended to estimate the random ε's, which cannot be observed. Residuals are usually standardized by dividing them by their standard errors. The result is called a "studentized" residual, because it has a Student's t distribution.

$$s_i = \frac{y_i - \hat{y}_i}{\text{std. error}(y_i - \hat{y}_i)} \tag{5.38}$$

A plot of studentized residuals versus predicted values is used to detect relationships between the estimates of the errors and the estimated size of the response. Additionally, plots of studentized residuals versus the independent variables are used to check for linearity, and plots of studentized residuals versus normal scores are used to check for normality of errors.

A plot of studentized residuals versus predicted values from the final model for Y_1 in the Surface example is given in Fig. 5.10. The plot shows evidence that the variance of the residuals is greater for predicted values farther from the middle of the data, an indication of violation of the assumption of homogeneity of variance. Although the relationship does not appear to be strong enough to warrant changing the model, a researcher should use prediction intervals based on this model with caution. Prediction intervals for points near the edges of the X data might be too narrow because actual variance of errors in that area is larger than average.

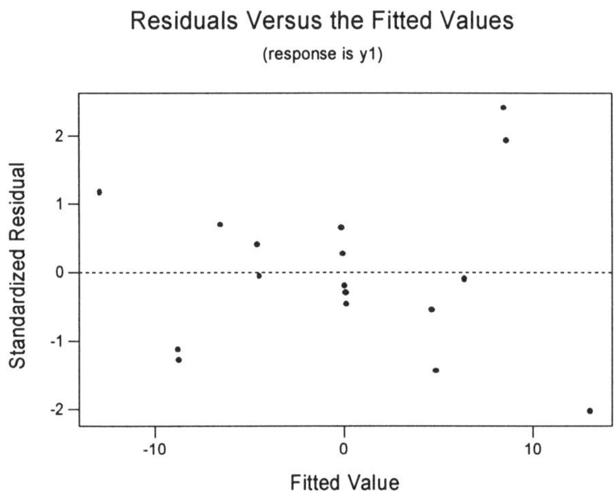

Fig. 5.10. Studentized residuals versus predicted values for the final model for Y_1 in the Surface data.

A common pattern in plots of studentized residuals versus predicted values is a "funnel" shape, in which variance of residuals increases with the predicted value. Such a pattern, if strong, suggests that there is a systematic relationship between the size of the process and the variance of the process. Analysis on a transformed scale (e.g. log or square root) might have homogeneous variance. If analysis of transformed variables is unappealing, another option is a "weighted" least squares analysis (described later in the context of nonlinear regression). For moderate to large n, the distribution of the studentized residuals is approximately normal, so residuals well beyond the range –2 to 2 might be suspected of being errors or anomalies (called "outliers").

To demonstrate the use of plots of studentized residuals versus individual X variables, we inspect residual plots from the model in which Y_1 is predicted using X_1 and

X_2, (without squared or cross-product terms). The plot of studentized residuals versus predicted values (Fig. 5.11) indicates clear evidence of curvature that is also evident in the plot of studentized residuals versus X_1 (Fig. 5.12). There is no evidence of curvature in the plot of studentized residuals versus X_2 (not shown). These results are not surprising, since the analysis has already demonstrated a need for an X_1^2 term in the model, but not a need for an X_2^2 term.

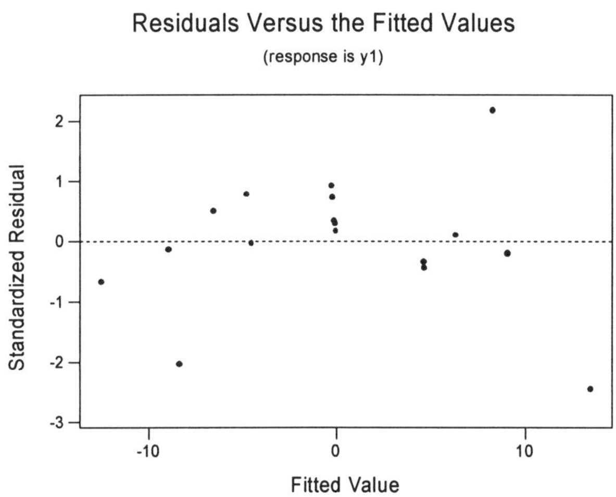

Fig. 5.11. Studentized residuals versus predicted values for the regression of Y_1 on X_1 and X_2 in the Surface data.

5.4 NON-LINEAR REGRESSION

5.4.1 The Radon Build-up example

One method for determining the concentration of ^{226}Ra in environmental samples is to take advantage of serial radioactive decay, where the buildup of ^{222}Rn in a sealed sample is measured as a function of time. The method actually measures the 609 keV gamma rays from the radon daughter ^{214}Bi. The model for the case where the sample is sealed at time $t = 0$ is:

$$y = A_0 e^{-\lambda t} + A_\infty (1 - e^{-\lambda t}) + \varepsilon . \tag{5.39}$$

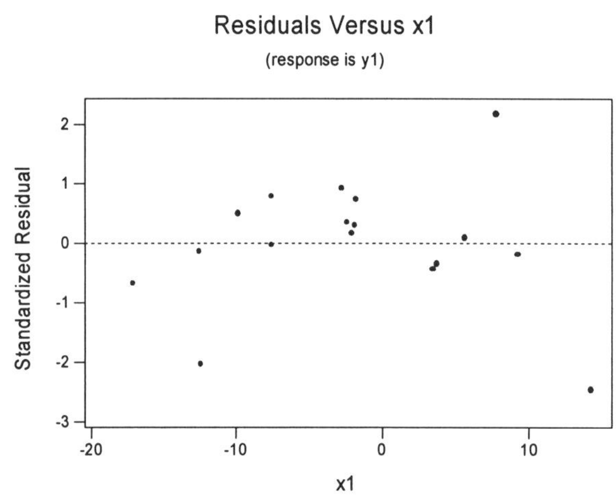

Fig. 5.12. Studentized residuals versus X_1 for the regression of Y_1 on X_1 and X_2 in the Surface data.

Here A_0 represents ^{222}Rn that has not escaped from the sample during collection and preparation. A_∞ is the total ^{222}Rn that will accumulate in the sealed sample when t is large compared with the half-life. A_∞ is in effect equivalent to the activity of ^{226}Ra at equilibrium. The primary objective is to determine A_∞ . The researcher may also solve for the half-life, $T_{1/2} = \ln(2)/\lambda$. The quantity $(1 - A_0 / A_\infty)$ is also of interest and is referred to as the radon emanation factor.

The easiest way to do this would be to wait until the ^{222}Rn comes into equilibrium with the ^{226}Ra and make a single measurement of the activity in the sample. This would require a delay of about one month, which is often undesirable or unacceptable. Another approach would be to make several measurements of y_i at time points, t_i and use regression techniques to solve for the coefficients in the model, namely, A_0, A_∞ and λ.

Rewriting Eqn. 5.39 gives the following expression:

$$y = A_\infty + (A_0 - A_\infty)e^{-\lambda t} + \varepsilon . \tag{5.40}$$

If the decay rate were known, then the term $e^{-\lambda t_i}$ could be calculated, and A_∞ and (A_0 − A_∞) could be estimated using the slope and intercept of a simple linear regression of y_i on $e^{-\lambda t_i}$. The estimate of A_0 is then obtained by adding the slope and intercept. One reason *not* to assume that λ is known is to see if the estimate of λ agrees with the known value for ^{222}Rn. There might be contamination from other radionuclides that might alter the estimated value of λ.

If A_∞ were known, then (y_i - A_∞) would be known, and Eqn. 5.40 would be "linearizeable"; that is, it could be made linear by subtracting A_∞ from both sides then taking logarithms (base *e*):

$$\ln(y_i - A_\infty) = \ln(A_0 - A_\infty) - \lambda t_i \tag{5.41}$$

The slope of a simple linear regression of $\ln(y_i - A_\infty)$ on t_i gives an estimate of -λ, and the intercept provides an estimate of $\ln(A_0 - A_\infty)$, from which (using the known value of A_0) an estimate of A_∞ can be calculated. It is only when both A_∞ and λ are unknown that the model is inherently nonlinear.

Counts are observed at n = 15 time points as given in Table 5.2 and plotted in Figure 5.13. We fit initially a nonlinear regression model to all 15 points. We will then examine how the estimate and confidence interval for A_∞ change when the regression is based on fewer points.

Table 5.2. Time (hours) and Count (per min.) for the Radon Build-up example.

Time (h)	Count (per min.)	Time (h)	Count (per min.)
3	1117.486	69	7303.440
6	1439.501	75	7534.774
9	1779.421	93	8521.968
21	3201.219	99	8980.658
27	3634.399	117	9801.010
33	4210.212	141	11096.100
45	5241.435	165	12017.590
51	5668.028		

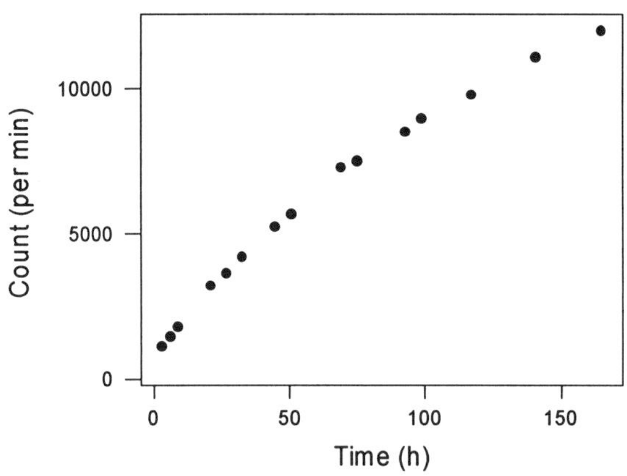

Fig 5.13. Time (hours) versus Count (per min.) for the Radon Build-up example.

5.4.2 Least squares estimation of parameters

Least squares estimation of parameters for a non-linear model follows the same basic formula as for a linear model. Parameter estimates are found that minimize the quantity:

$$SSE = \sum_{i=1}^{n}\left(y_i - (A_\infty + (A_0 - A_\infty)e^{-\lambda t_i})\right)^2 . \tag{5.42}$$

Most statistical packages offer the choice of several methods to minimize *SSE*. Some of the methods require that the user supply derivatives of *SSE* with respect to the parameters. In general, minimization requires an iterative procedure that may converge to a local (rather than a global) minimum, or converge to parameter estimates outside the parameter space (in our case negative estimates), or not converge at all. Some care may be needed adjusting convergence criteria of the minimization algorithm. Minimization is usually less troublesome if the user can supply a reasonable set of parameter starting values for the first iteration.

Finding starting values is a somewhat *ad hoc* procedure. In the Radon Build-up data, A_0 is the intercept of the curve and can be estimated by eye as approximately 800. Since A_∞ is the limiting height of the curve as time increases, we may guess it to be about 25,000. Putting those initial estimates for A_0 and A_∞ into Eqn. 5.40, we obtain:

$$y = 25000 + (25000 - 800)e^{-\lambda t} . \tag{5.43}$$

Subtract 25,000 from each side and take logarithms to obtain an equation that is linear in λ:

$$\ln(y_i - 25000) = \ln(25000 - 800) - \lambda t_i . \tag{5.44}$$

A simple linear regression of $\ln(y_i - 25000)$ on t_i yields a slope of –0.0038, which suggests an initial value of λ=0.0038. To check these initial values, predicted values using those estimates are computed and plotted along with the data in Fig. 5.14.

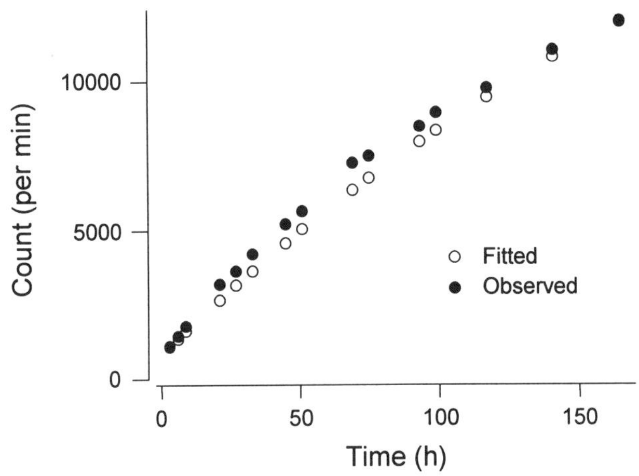

Fig 5.14. Plot of predicted count values using the initial estimates for the Radon Build-up data.

The appearance of the plot indicates that the guess of 25,000 for A_∞ is probably too high, which has led us to an initial guess for λ that is probably too low. Rather than try to improve these starting values, we try them in SAS Proc Nlin (SAS Institute, 1996). The SAS commands used and portions of the SAS output are given in Exhibit 5.9. Since no model derivatives were specified, SAS will compute approximate derivative values from a history of iterations.

After seven iterations, the algorithm converged to the least squares parameter estimates $\hat{A}_\infty = 16{,}797.5, \hat{A}_0 = 800.6$ and $\hat{\lambda} = 0.00725$. Since *MSE* = 9689.0, $\hat{\sigma} = \sqrt{9689} = 98.4$. It is a good idea to check the fit by again plotting the predicted values along with the data. The fit, given in Fig. 5.15, looks very good.

Exhibit 5.9. SAS commands and abbreviated output for the first run of the radon build-up data.

Commands:
```
data buildup;
infile 'c:\My Documents\buildup.txt';
input time count;
proc nlin;
        model count=ainf+(azero-ainf)*exp(-lam*time);
        parms ainf=25000 azero=800 lam=0.038;
run;
```
Output:
```
                              The NLIN Procedure
                               Iterative Phase
                          Dependent Variable count
                            Method: Gauss-Newton

                                                              Sum of
          Iter          ainf          azero          lam      Squares
             0       25000.0          800.0      0.00380      3820620
             1       23064.5          803.6      0.00421      3484365
             2       20686.1          808.0      0.00486      3159948
             3       18865.7          810.2      0.00558      2366853
             4       16758.6          809.1      0.00683      1126030
             5       16771.1          801.9      0.00724       118861
             6       16797.6          800.7      0.00725       116267
             7       16797.5          800.6      0.00725       116267
    NOTE: Convergence criterion met.

                                 Sum of        Mean    Asymptotic     Approx
Source                   DF     Squares      Square       F Value     Pr > F
Regression                3    7.3415E8    2.4472E8       9052.64     <.0001
Residual                 12      116267      9689.0
Uncorrected Total        15    7.3426E8
Corrected Total          14    1.7554E8

                                 Asymptotic
                                   Standard     Asymptotic 95% Confidence
        Parameter    Estimate         Error                Limits
        ainf          16797.5         472.2       15768.6        17826.4
        azero           800.6       62.5649         664.3          937.0
        lam           0.00725      0.000353       0.00648        0.00802
```

There are several versions of R^2 that are used for nonlinear regression. The most easily interpretable version is analogous to the R^2 of linear regression, calculated as one minus the proportion of variability unexplained:

$$R^2 = 1 - \frac{SSE}{SST} = 1 - \frac{116267}{1.75540000} = 0.9993 \tag{5.45}$$

Thus, 99.93% of the variability in Count has been explained by the nonlinear model in Time. A common mistake is to calculate R^2 as the ratio of Regression SS (*SSR)* to Uncorrected *SST.* This ratio overstates the quality of the regression because it uses variability in Count about zero, rather than the more commonly used variability about the mean. The same problem would occur in linear regression when the intercept term is omitted from the model. An alternate version of R^2 is computed as the square of the Pearson product moment correlation, *r,* between the observed and predicted values. It has been noted that in linear regression, that the two definitions are equivalent. In nonlinear regression, they are not.

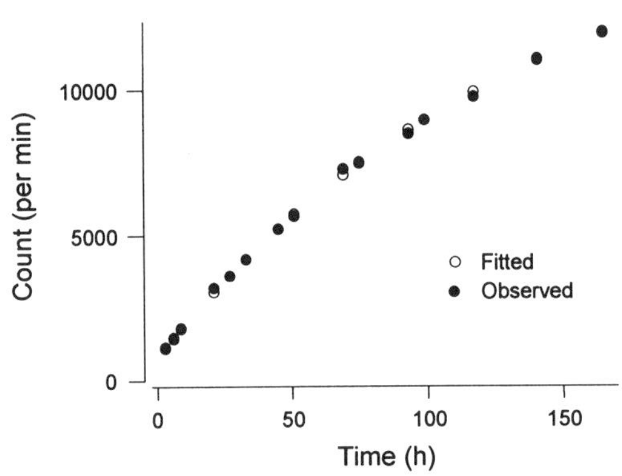

Fig 5.15. Plot of predicted count values using the least squares estimates for the Radon Build-up data.

Sometimes initial values are not easy to find. When that is the case SAS will accept a grid of parameters, which it will search to find suitable starting values. Replace the "parms" statement in the SAS code by:

```
parms ainf=15000 to 30000 by 1000
      a0=800
      lam=0.0038 to 0.01 by 0.001;
```

This directs SAS to fit the model for all combinations of parameters specified, and then begin the iterative minimization procedure from the combination of parameter values with the lowest *SSE*. (In our example, requesting the above grid search yielded the same final result achieved using our crude starting values.)

SAS computes "Asymptotic standard errors" by a method that assumes that the parameter estimates are approximately linear functions of the data. The accuracy of that assumption improves as n approaches infinity (hence the term "asymptotic). "Asymptotic 95% Confidence Limits" are computed as the estimate plus and minus the standard error multiplied by the appropriate value from the student's t table. Hypothesis tests based on approximate T-statistics can be constructed, but are not automatically computed by SAS. Since the $n = 15$ in the Radon Build-up example is not very close to infinity, these standard errors, and T-tests and confidence intervals should be taken as very approximate. That is, the actual coverage rate of confidence intervals based on this method may be somewhat different than the claimed value of 95%.

5.4.3 Weighted least squares estimation of parameters

The validity of any hypothesis test or confidence interval depends on the validity of the assumptions of the model. In this example we have reason to question the assumption of homogeneity of variance. The usual model for radiation counts is the Poisson distribution, which has variance equal to its mean. Since the regression line is, by definition, the mean, the variance of the errors might be assumed to be proportional to the height of the regression line. Since the estimated curve is believed to be approximately accurate, then the count at 165 minutes is approximately ten times more variable than the count at 3 minutes. Statistical theory indicates that least squares estimates are most efficient when *SSE* is computed using weights inversely proportional to the variance. The "weighted least squares" estimates are the parameter estimates that minimize:

$$SSE = \sum_{i=1}^{n}\left(\frac{1}{\mathrm{var}(\varepsilon_i)}(y_i - (A_0 + (A_1 - A_0)e^{-\lambda t_i}))^2\right). \qquad (5.46)$$

Since var(ε_i) is proportional to the height of the line, we may take the height of the line estimated using the (ordinary) least squares estimates as an adequate approximation. The effect of weighting is typically to produce an estimated curve that more closely fits the data points that have lower y-values, at the expense of a poorer fit to the data points that have higher y-values.

Proc Nlin allows a weighted least squares option. In Exhibit 5.10 SAS commands are given that create a new data set (buildup2) in which weights are computed based on first (ordinary) least squares. The weights are then used in the subsequent Proc Nlin. (In the economics literature this procedure is called "two-stage least squares".) The final Proc Nlin demonstrates a feature in Proc Nlin that computes weighted least squares estimates, with the weights updated at each iteration of the minimization routine. This is called "iteratively re-weighted least squares". The parameter estimates for the two procedures are given at the bottom of the table.

The estimates and confidence intervals for the two weighted least squares estimates are very close to each other. However, these estimates are different from the un-weighted (called "ordinary") least squares estimates of the previous section. In particular, the weighted least squares A_0 parameter estimates are somewhat lower than the ordinary least squares estimate, which is 16797.5. Also, the confidence intervals for A_0 from the weighted least squares analyses are wider than the un-weighted analysis (2092 versus 2058), reflecting the greater uncertainty about the limiting value when the higher counts of the later time points are assumed to be more variable. Because the confidence intervals from the weighted least squares analysis account for the unequal error variance, confidence interval coverage may be closer to the nominal value of 95%.

It is of interest to compare the estimated half-life based on the estimated value of λ with the known half-life for ^{222}Rn of 3.82 days.

$$\hat{T}_{1/2} = \frac{\ln(2)}{\hat{\lambda}} \times \frac{1}{24\text{hrs./day}} = \frac{0.693}{0.00748} \times \frac{1}{24} = 3.86 \text{ days} \qquad (5.47)$$

The estimate is very close to the theoretical value, giving no evidence of contamination of the sample.

Exhibit 5.10.

<u>SAS commands:</u>

```
data buildup2; * A little program to compute weights;
     set buildup;
     ainf=16797.5;
     azero=800.6;
     lam=0.00725;
     yhat=ainf+(azero-ainf)*exp(-lam*time);
     wt=1/yhat;
proc nlin;  *A WLS using 1/yhat from above as weights;
   model count=ainf+(azero-ainf)*exp(-lam*time);
   parms ainf=16797.5
         azero=800.6
         lam=0.00725;
    _weight_=wt; * Uses wt calculated above;
proc nlin; *Demonstrates Nlin's interatively reweighted l.s. ;
   model count=ainf+(azero-ainf)*exp(-lam*time);
   parms ainf=16796.5
         azero=800.6
         lam=0.00725;
   _weight_=1/model.count;  *Reweights at each iteration;
run;
```

<u>Two-stage least squares estimates from SAS output:</u>

Parameter	Estimate	Asymptotic Standard Error	Asymptotic 95% Confidence Limits	
ainf	16512.5	480.8	15464.9	17560.0
azero	771.5	34.1593	697.1	846.0
lam	0.00748	0.000349	0.00672	0.00824

<u>Iteratively re-weighted least squares estimates from SAS output:</u>

Parameter	Estimate	Asymptotic Standard Error	Asymptotic 95% Confidence Limits	
ainf	16512.3	480.1	15466.4	17558.3
azero	771.4	33.9032	697.5	845.2
lam	0.00748	0.000348	0.00672	0.00824

5.4.4 Estimates based on fewer time points

Observation of the 15th time point requires a wait of 165 hours (almost 7 days). We now consider estimates of A_∞ based on fewer than the full 15 observations and compare their accuracy to the estimate based on n = 15. See Table 5.3. Of course, the advantage of using fewer time points is a shorter wait for the estimate, and the disadvantage is diminished accuracy of the estimate, which will be reflected in wider confidence intervals.

Table 5.3. Estimates and confidence intervals for A_∞ in the Radon Build-up data using fewer than the available number of points.

Data pts	15	13	11	9	7	5
Upper C.L.	17558	16999	18232	25689	19514	56364
A_0 Estimate	16512	15459	15452	17701	12342	12030
Lower C.L.	15466	13919	12672	9712	5170	-32304
C.I. Width	2092	3080	5560	15977	14344	88668

The estimates of A_∞ become erratic with fewer than 11 data points, and the confidence intervals become much wider. However, the validity of these confidence intervals is based on large-sample assumptions, which were questioned, even when the estimates were based on n = 15 points. It is hard to know at which sample size to quit believing that the intervals have any value at all. Clearly, the results for n = 5 are invalid, because the interval is not only extremely large, but includes negative numbers.

5.4.5 Concluding remarks about nonlinear regression

There are many topics and extensions in nonlinear model fitting that are beyond the scope of this simple introduction. All of the issues that arise in linear regression modeling also arise in nonlinear regression modeling, and more! For example, in nonlinear regression modeling, the accuracy of the asymptotic confidence intervals can change, depending on a choice of parameter specification. A model may use an equation that contains the value e^β. Say the model is rewritten, replacing e^β by the single parameter φ. In principle, it should make no difference whether the model is written in terms of β or φ. But the accuracy of asymptotic confidence intervals may be good for φ, but poor for β, or vice versa. This is a problem that goes by the title "parameter effects curvature" that does not occur in linear regression. Although such issues are beyond the scope of this presentation, a reader should be able to use the information presented here to get at least a start on simple nonlinear modeling problems.

5.5 REFERENCES

Kim, D. S; Kim, Y. S. Distributions of airborne radon concentration in Seoul metropolitan subway stations. Health Phys. 65:12-16; 1993.

MINITAB, Inc. Meet MINITAB, Release 12 for Windows, State College, PA: MINITAB, Inc., 1997.

SAS Institute, Inc. SAS/STAT User's Guide, Version 6, Fourth Edition, Volume 2, Cary, NC: SAS Institute Inc., 1989.

5.6 AUTHOR PROFILE

Dr. Phillip Chapman is Associate Professor of Statistics and Experimental Station Statistician at Colorado State University. He has a B.A. in mathematics from the University of California at Berkley, and a Ph.D. in statistics from the University of Minnesota. He has 16 years experience teaching statistics, and consulting with university and industry researchers, primarily in the agricultural, biological, environmental and biomedical sciences. Dr. Chapman teaches applied statistical methods to graduate students and researchers throughout Colorado State University. His research interests include applied linear models, environmental statistics, and bootstrap methods.

Chapter 6

NONPARAMETRIC STATISTICAL METHODS

Hari K. Iyer

6.1 INTRODUCTION

6.1.1 Parametric Versus Nonparametric Statistics

Statistical inference procedures are often classified as parametric methods or as nonparametric methods. Parametric methods are statistical inference tools that may be used when data are adequately modeled using *parametric families* of distributions. These are distributions whose shapes are completely determined by a few population summary characteristics -- typically the population mean and the population standard deviation. The family of normal distributions is perhaps the most commonly used parametric family; other commonly used parametric families are the lognormal, gamma, binomial, and Poisson distributions. Most introductory statistics courses cover statistical methods that assume the data are normally distributed.

Nonparametric methods are statistical inference tools that may be used when it is not known whether data may be adequately modeled using known parametric families of distributions. Generally, it is assumed only that the individual data values are *independent*. Sometimes it is assumed that the data distribution is *continuous,* i.e., measured on a continuum. Sometimes it is assumed that the data distribution is *symmetric,* but it is not assumed that the data belong to a parametric family.

The purpose of this chapter is to provide a suite of nonparametric statistical tools that are widely applied to make decisions about the environment we live in. It is assumed that the reader is somewhat familiar with standard parametric procedures such as t-tests, confidence intervals , and straight-line regression. Each of these parametric procedures typically has at least one nonparametric counterpart. In the following subsection we discuss some of the issues that need to be considered when choosing between a parametric method and a nonparametric method for a particular application.

6.1.2 When Would One Use Nonparametric Methods?

In many disciplines, past experience might suggest that certain types of parametric models generally provide adequate fit to the data and inference procedures based on such models provide valid answers. An example is the use of the lognormal distribution for modeling environmental data. Sometimes, especially when large amounts of data are available, one can find a suitable standard parametric distribution for modeling it. By standard parametric family of distributions we mean distributions for which statistical inference procedures are well developed; this includes the normal distribution, the gamma distribution, the Poisson distribution, and the binomial distribution. In situations where neither past experience nor the data themselves provide guidance for selecting a parametric model it is appropriate to consider nonparametric methods.

A natural question to ask then is, "why would one not use nonparametric methods all the time?" since these methods do not rely on questionable distributional assumptions. If a suitable parametric model is available for describing a particular situation then confidence intervals based on the parametric model will be narrower and hypothesis tests will have more *power* for detecting effects or changes that are being investigated. If a parametric model is appropriate then parametric procedures are more "efficient." One way to think about this is that, with a fixed amount of data, more information can be extracted using appropriate parametric methods than with nonparametric methods. Only when we have no confidence in the choice of a parametric method would we consider a nonparametric method for answering a question.

The above discussion leads to the question "what can go wrong if one were to use parametric methods all the time even though the required underlying assumptions may not be satisfied?" Only when a parametric model is a good approximation to the data distribution can we trust the probability calculations used to quantify uncertainty in the statistical analyses. The actual confidence levels will indeed be close to the claimed confidence levels (e.g. 95%) and the actual type-I and type-II error rates for hypothesis tests will be close to the claimed error rates only when the distributional assumptions are valid. When the assumed parametric model is grossly incorrect, these uncertainty estimates will also be grossly incorrect. A 95% confidence interval may have an actual coverage that is substantially below 95%. A hypothesis test with alpha equal to 0.05 may have an actual type-I error rate that is much larger than the claimed 5%, and so on.

We might ask the question "how will we know if the parametric assumptions are reasonable in a given situation?" There exist graphical techniques as well as hypothesis tests that provide guidance for judging the adequacy of an assumed model. These include probability plots, various goodness-of-fit tests, and tests of normality. The reader is referred to Conover (1999) for further information on these topics.

A common belief about nonparametric methods is that they do not require any assumptions. This is false. Nonparametric methods do rely on certain assumptions but these are less stringent than the assumptions required by parametric methods. Typically one needs to assume that the data are a random sample from some well-defined population. When the sample is not a planned random sample, one should check for presence of serial correlations and/or spatial correlations. This typically requires expert statistical assistance.

The rest of this chapter will discuss some commonly used nonparametric statistical methods and illustrate their applications using real-life examples. Although formulas will be presented for the sake of completeness, the calculations will be carried out using the statistical package MINITAB (1997) and the relevant portions of the computer output will be included. For information about MINITAB please consult the company website at http://www.minitab.com.

6.2 COMMON PARAMETRIC METHODS AND THEIR NONPARAMETRIC COUNTERPARTS

Table 6.1 gives a partial list of some of the standard parametric statistical methods along with their nonparametric counterparts. Typically, there is at least one nonparametric technique corresponding to each popular parametric technique. In this chapter, we will discuss almost all of the nonparametric techniques listed in Table 6.1. It is assumed that readers are somewhat familiar with some or all of the parametric methods listed.

6.3 EXPLORATORY DATA ANALYSIS

Exploratory data analysis is an essential first step in getting to know the data. This consists of graphical representations of the data and calculation of summary statistics. A *statistic* is a function of the data that numerically describes some characteristic of the data. Some well-known examples of summary statistics are -- mean, median, percentiles, standard deviation, range, inter-quartile range, minimum, and maximum. These summary statistics give us clues about populations from which the data

Table 6.1. Some common parametric statistical methods and their nonparametric counterparts

Parametric	Nonparametric
Summary Statistics: Mean, Standard deviation, etc.	Summary Statistics: Median, percentiles, Inter-quartile range
Proportions of data above or below given thresholds. Tests and confidence intervals.	Proportions of data above or below given thresholds. Tests and confidence intervals. (same as in the parametric case)
Single sample t-test	Sign test; Signed Rank Test
Two sample t-test	Mann-Whitney test
Paired t-test	Wilcoxon's signed ranks test
k-sample F-test	Kruskal-Wallis k-sample test
Randomized complete block design	Friedman's test
Pearson correlation	Spearman correlation
Least Squares Estimation of slope & intercept in a straight-line model	Sen's method of estimation of slope in a straight-line model

were sampled. This in turn allows us to judge whether or not a parametric technique is suitable for analyzing the data.

Graphical displays of datasets facilitate discovery of relationships and patterns among the variables included in the dataset. They also help us identify peculiar data values and/or other problems with the data; additionally, they can help us decide whether or not assumptions needed to justify a chosen statistical procedure are met. For instance, a normal probability plot can help us decide whether the assumption of normality of the population distribution is reasonable. They can also help us discover outliers.

6.3.1 Example: Distribution of ^{222}Rn in Groundwater

Data were collected as part of the Apalachicola-Chattahoochee-Flint River Basin NAWQA (National Water Quality Assessment) Study [*ACF River Basin Study* for short]. The histogram in Fig. 6.1 shows the distribution of ^{222}Rn in the samples. The normal distribution curve that has the same mean and standard deviation as the sample data is superimposed on the histogram to help us visually judge the adequacy of a normal distribution model for these data. It is quite clear here that the normality assumption is violated.

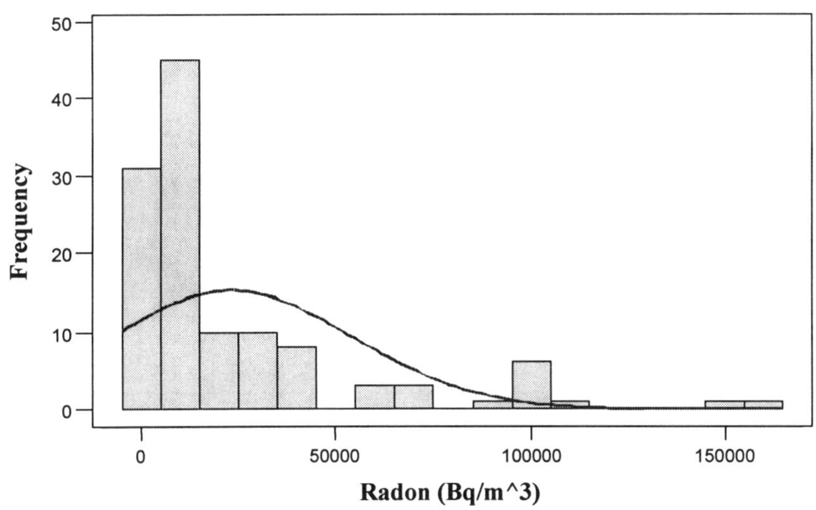

Fig. 6.1. Histogram of Radon data (Bq/m^3) with superimposed normal curve

MINITAB can be used to produce *probability plots* such as the ones shown in Fig. 6.2. These plots can help us decide whether a parametric model is available for describing the data distribution. Each graph in Fig. 6.2 corresponds to a particular distributional family. The distributions included are the normal, the lognormal, the exponential, and the Weibull. Each plot uses a special scaling for the vertical axis determined by the percentiles of the particular distribution being considered. In some cases the horizontal axis also uses special scaling, e.g., lognormal and Weibull. The scales are determined in such a way that, if the distribution under consideration is a good model for the data, then the plotted points will fall roughly on the straight line exhibited on the graph, implying that there is good agreement between the percentiles determined from the data and the percentiles determined from the theoretical distribution. Based on the graphs in Fig. 6.2, the lognormal distribution appears to provide a reasonable model for the ^{222}Rn data. Therefore one might wish to use parametric inference methods that have been developed for lognormal distributions.

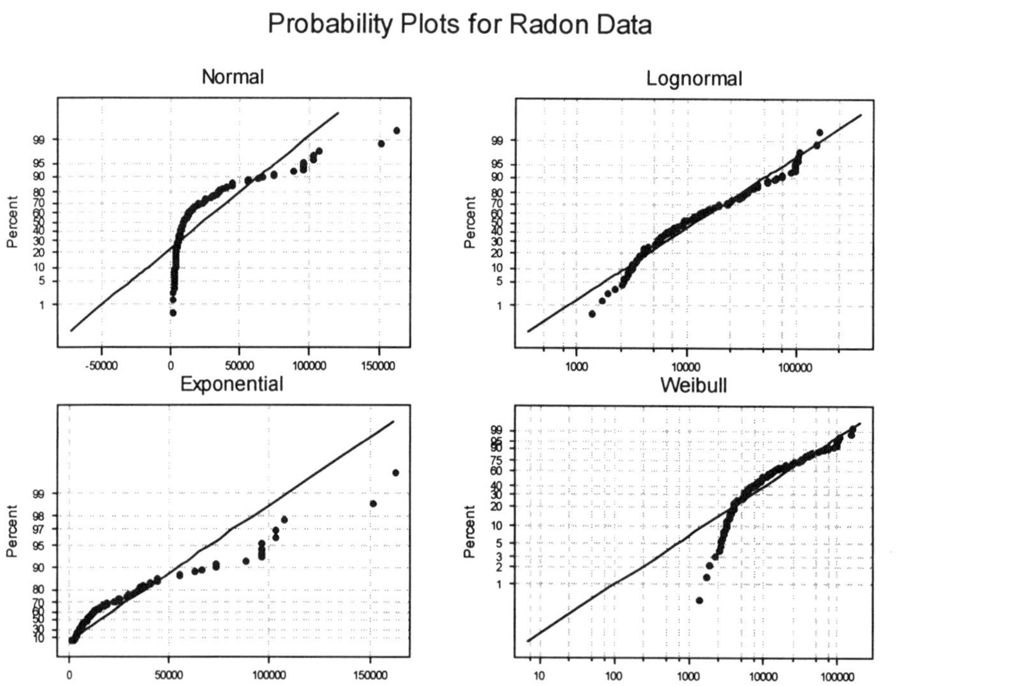

Fig. 6.2. Probability plots for the Radon data. The distributions being considered are the Normal, the Lognormal, the Exponential, and the Weibull.

The ACF River Basin Study also provided data on Tritium concentrations. A histogram for these data is shown in Fig. 6.3 with a superimposed "best fit" normal curve. Again it is seen that the normal distribution gives a very poor fit. Fig. 6.4 shows the probability plots for the Tritium data using the normal, lognormal, exponential, and the Weibull distributions. None of these distribution types appears to provide a good model. So one would consider a nonparametric method of analysis for these data.

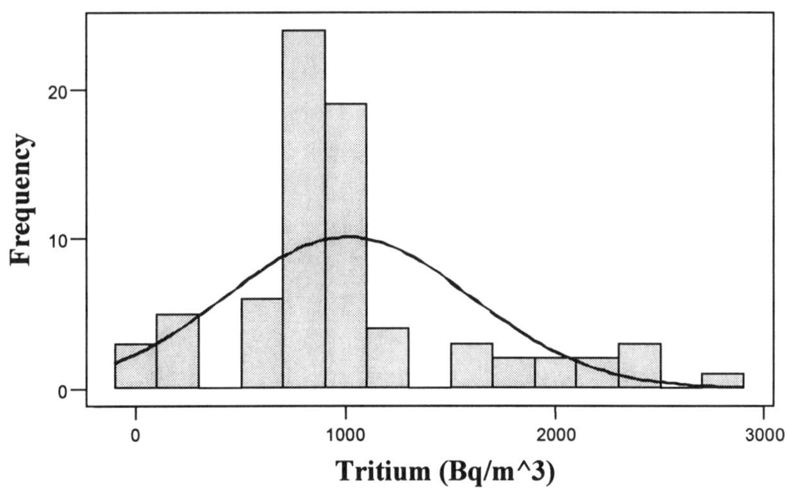

Fig. 6.3. Histogram of Tritium data (Bq/m^3) with superimposed normal curve

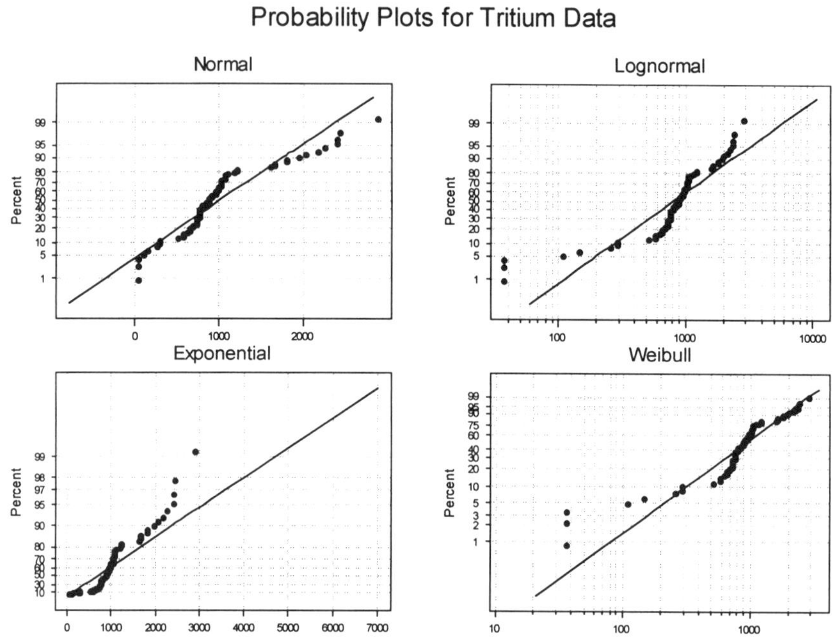

Fig. 6.4. Probability plots for the Tritium data. The distributions being considered are the Normal, the Lognormal, the Exponential, and the Weibull.

6.3.2 Numerical and Graphical Summaries

The display shown in Fig. 6.5, obtained using MINITAB, provides descriptive statistics for the ^{222}Rn data. Reported values include the mean and the standard deviation of the data and a FIVE NUMBER SUMMARY – minimum, 1st quartile, median, 3rd quartile, and the maximum. Recall that 1st quartile refers to the 25th percentile and 3rd quartile refers to the 75th percentile. These are sometimes denoted using the symbols Q1 (first quartile) and Q3 (third quartile), respectively. The difference (Maximum – Minimum) is called the Range and the difference (Q3-Q1) is called the Interquartile Range. The standard deviation, the range, and the interquartile range describe the variability present in the data whereas the mean and the median attempt to direct our attention to the center of the data distribution.

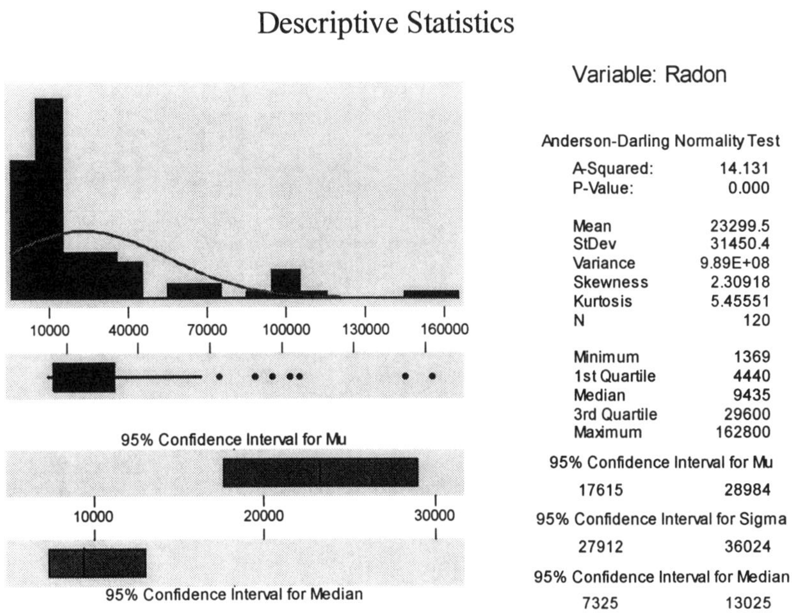

Fig. 6.5. Numerical and graphical summaries for Radon data using MINITAB

Fig. 6.5 also includes a box plot of the data and confidence intervals for the population mean, population standard deviation, and the population median. The points in the box plot, shown as filled in circular symbols, are "potential outliers", and the validity of these data values should be investigated prior to any data analysis. The distribution of ^{222}Rn values is seen to be highly skewed. This is a reason why the mean and the median of the data are so different from each other. For symmetric data distributions the mean and the median will coincide. Summary information, along with graphs, for Tritium data is shown in Fig. 6.6.

Descriptive Statistics

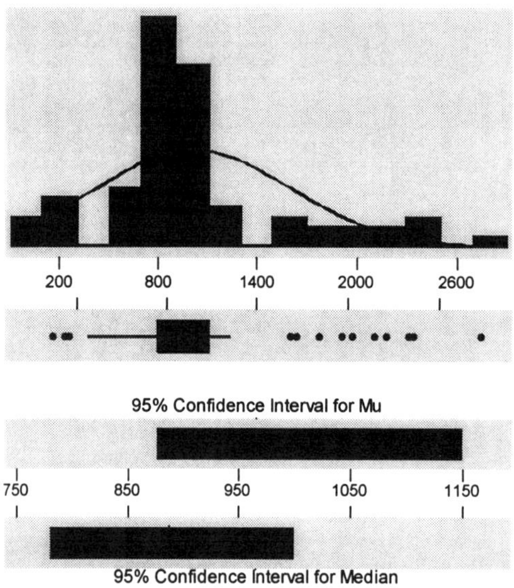

Variable: Tritium

Anderson-Darling Normality Test

A-Squared:	3.724
P-Value:	0.000
Mean	1012.50
StDev	587.73
Variance	345422
Skewness	1.11983
Kurtosis	1.42993
N	74
Minimum	37.00
1st Quartile	740.00
Median	888.00
3rd Quartile	1073.00
Maximum	2886.00

95% Confidence Interval for Mu

876.33 1148.67

95% Confidence Interval for Sigma

505.92 701.34

95% Confidence Interval for Median

780.15 999.00

Fig. 6.6. Numerical and graphical summaries for Tritium data using MINITAB

6.4 THE SIGN TEST FOR A MEDIAN

A common statistical problem that occurs in environmental data analyses is to decide, based on data, whether or not the mean level of contamination exceeds a reference value, which we denote by x_0. The standard parametric approach to this problem uses a t-test. The nonparametric formulation of this problem considers the median rather than the mean. The hypotheses may take one of three forms.

6.4.1 Left-Sided Alternative

In the case where the objective is to demonstrate that the population median is smaller than a reference value x_0, the null and the alternative hypotheses may be stated as follows.

Null Hypothesis: The population median concentration is greater than or equal to x_0

Alternative Hypothesis: The population median concentration is less than x_0

This is the form of the test needed to demonstrate compliance with a regulatory standard.

6.4.2 Right-Sided Alternative

In the case where the objective is to demonstrate that the population median is greater than a reference value x_0, the null and the alternative hypotheses may be stated as follows.

Null Hypothesis: The population median concentration is less than or equal to x_0

Alternative Hypothesis: The population median concentration is greater than x_0

This is the form of the test needed to demonstrate that compliance with a regulatory standard is being violated.

6.4.3 Two-Sided Alternative

In the case where the objective is to demonstrate that the population median is different from a reference value x_0, the null and the alternative hypotheses may be stated as follows.

Null Hypothesis: The population median concentration is equal to x_0

Alternative Hypothesis: The population median concentration is not equal to x_0

The Sign test is a nonparametric procedure that allows us to test these hypotheses under the assumption that the sample data are randomly selected from the population of interest. The population distribution is assumed to be such that the proportion of data values strictly below the median is equal to the proportion of data values strictly above the median. The computational details for the sign test are given below.

6.4.4 Computations for the Sign Test

Let the n data values be denoted by $X_1, X_2, \cdots, X_n$. It is assumed that these n values are a random sample from the population of interest. Remove all data values that are exactly equal to x_0. Let the number of remaining data values be denoted by m. We will refer to m as the effective sample size for the sign test. The decision rules under each of the three scenarios described above – left sided alternative, right-sided alternative, and two-sided alternative – are given below.

6.4.4.1 Left-Sided Alternative. Let T equal the number of data values strictly greater than x_0. Clearly, small values of T support the alternative hypothesis. The rule is to reject the null hypothesis if T is strictly less than a suitably chosen value T^*, called a *critical value*. T^* is obtained from an appropriate table or a computer program. Its value depends on the sample size and on the type-I error rate (probability of rejecting a true null hypothesis) that is selected for the given application. This error rate is generally denoted by the symbol α and is a number between 0 and 1. Commonly used values for α are 0.05 or 0.01. Table-A at the end of the chapter gives the critical values of T for selected values of α, for effective sample sizes ranging from 5 to 150. The value of T^* is the entry for α in the row corresponding to m (effective sample size). For larger sample sizes it is suggested that one use a suitable computer program such as MINITAB to carry out the test. Because the test statistic T is a count and can take only integer values, not every α level is realizable, the critical values given in Table-A are computed so that the type I error rate is as close as possible to the specified value α without exceeding it.

6.4.4.2 Right-Sided Alternative. Let T equal the number of data values strictly less than x_0. Again, small values of T support the alternative hypothesis. Therefore, the rule is to reject the null hypothesis if T is strictly less than the critical value T^* from Table-A at the end of the chapter. As before, T^* is the entry for α in the row corresponding to m (effective sample size).

6.4.4.3 Two-Sided Alternative. Let T_1 denote the number of data values strictly greater than x_0 and T_2 denote the number of data values strictly less than x_0. Reject the null hypothesis if either T_1 is strictly less than T^* or T_2 is strictly less than T^* where T^* is the entry for $\alpha/2$ in Table-A, in the row corresponding to m (effective sample size). The reason for using $\alpha/2$ when determining the value of T^* in the two-tailed test is that, under a true null hypothesis the probability will then be (approximately) $\alpha/2$ that T^* will exceed T_1 and (approximately) $\alpha/2$ that T^* will exceed T_2. This will guarantee that the overall type-I error rate is (approximately) equal to the desired value α without exceeding it.

6.4.5 Sign Test for Radon Data

We illustrate the computations using the Radon data from the previous section. Suppose our objective is to demonstrate that the true median value for ^{222}Rn is less than 11,100 Bq/m (EPA standard for ground water). This corresponds to a test with a left-sided alternative with x_0 = 11,100. The test statistic T is equal to the number of data values greater than 11,100. This number is T = 55. The total number of data values is n = 120. There is one data value that is exactly equal to 11,100, so the effective sample size is m = 119. Using Table-A we find that the critical value for T corresponding to α = 0.05 is T^* = 51. Since T = 55 is not less than the critical value, the null hypothesis cannot be rejected.

Often the result of a statistical test is reported using a P-value, also called a significance probability. This is the smallest value of α for which the null hypothesis can be rejected based on the evidence in the data. P-value is a superior way to report the results of a statistical test. It measures the amount of evidence in the data to support the alternative hypothesis. The closer a P-value is to zero, the stronger is the evidence in the data to support the alternative hypothesis. Calculation of P-values generally requires a computer program.

Exhibit 6.1. Sign test for radon data

Sign Test for Median

Sign test of median = 11100 versus < 11100

	N	Below	Equal	Above	P	Median
Radon-222	120	64	1	55	0.2317	9435

Sign Confidence Interval

Sign confidence interval for median

	N	Median	Achieved Confidence	Confidence interval	Position
Radon-222	120	9435	0.9448	(7400, 12950)	50
			0.9500	(7325, 13025)	NLI
			0.9642	(7030, 13320)	49

Exhibit 6.1 gives the results of the sign test for the Radon data. It reports a P-value of 0.2317. Hence the null hypothesis cannot be rejected at α=0.05. It also reports a two-sided 95% confidence interval for the population median. This confidence interval covers the range 7,325 Bq/m^3 to 13,025 Bq/m^3. In particular, values above 11,110 Bq/m^3 are not ruled out. Thus, based on these data the null hypothesis that the median concentration of ^{222}Rn exceeds 11,100 Bq/m^3 cannot be rejected.

6.4.6 Sign Test for Tritium Data

Suppose the objective is to demonstrate, using α = 0.05, that the median Tritium concentration is greater than 740 Bq/m^3. Here x_0 = 740, n=74, and we are interested in a right-sided alternative. There are five data values exactly equal to 740, so the effective sample size is $m = 69$. The number of data values that are strictly less than 740 Bq/m^3 is T=16. Using Table-A with $m = 69$ and α = 0.05, we find that the critical value $T*$ is equal to 28. Since T= 16 is less than the critical value, the null hypothesis is rejected. Therefore the data support the statement that the true median concentration for tritium is greater than 740 Bq/m^3.

MINITAB output for the Tritium data is given in Exhibit 6.2. The P-value for the test is 0.0000 (zero to four decimal places). In particular, this means that the data provide very strong evidence in favor of the alternative hypothesis. Thus we can be quite certain that the median tritium value *exceeds* 740 Bq/m^3. A two-sided 95% confidence interval for the true median is from 780.1 Bq/m^3 to 999.0 Bq/m^3.

Exhibit 6.2. Sign test for tritium data

Sign Test for Median

Sign test of median = 740.0 versus > 740.0

	N	Below	Equal	Above	P	Median
Tritium	74	16	5	53	0.0000	888.0

Sign Confidence Interval

Sign confidence interval for median

	N	Median	Achieved Confidence	Confidence interval	Position
Tritium	74	888.0	0.9188	(814.0, 999.0)	30
			0.9500	(780.1, 999.0)	NLI
			0.9519	(777.0, 999.0)	29

6.5 THE WILCOXON SIGNED RANKS TEST FOR THE MEDIAN

If the population distribution is known to be approximately symmetric, then a more powerful nonparametric test is available for testing the population median. This is the Wilcoxon's signed rank test. Before using this test, it is advisable to check how well the symmetry assumption holds, by examining a histogram of the sample data. Bear in mind that sample histograms will not be perfectly symmetric even when the population distribution is symmetric. Judging sample histograms for symmetry requires some practice.

6.5.1 Symmetry plot

MINITAB provides a graphical tool, called the symmetry plot, which can be helpful in judging whether or not the data distribution is symmetric. A symmetry plot of sample data is constructed as follows. Arrange the data in increasing order and denote the ordered data by $X_{(1)}, X_{(2)}, \cdots, X_{(n)}$. Thus $X_{(1)}$ denotes the smallest data value, $X_{(2)}$ denotes the second smallest value, and so on. Calculate the median for the sample data. Denote this by M. For each positive integer i strictly less than $(n+1)/2$, calculate $U_i = X_{(n+1-i)} - M$ and $V_i = M - X_{(i)}$. If the number n of data values is odd, this process will result in $(n-1)/2$ ordered pairs; otherwise it will result in $n/2$ ordered pairs. A scatter plot of the pairs (U_i, V_i) with U_i on the horizontal axis and V_i on the vertical axis is called the symmetry plot of the sample data. On this plot, a reference line is drawn through the origin with slope equal to 1. If the data distribution is symmetric (at least approximately) then the points on this plot will be tightly clustered around the reference line. If most of the data lie below the reference line this indicates the data distribution is right-skewed (values above the median are much farther away from the median than values below the median). If most of the data lie above the reference line then the data distribution is left-skewed.

Fig. 6.7 shows a symmetry plot for the Tritium data and Fig. 6.8 is a symmetry plot for the square root transformed Tritium data. By default, Minitab superimposes a histogram on each symmetry plot. The data distribution in the original scale is right-skewed since the points are predominantly below the reference line. This is also clear from the histogram. The data distribution in the square root scale appears to be more symmetrically distributed than in the original scale.

The details needed to carry out the Wilcoxon Signed-Ranks test are discussed next.

6.5.2 The Wilcoxon Signed Ranks Test -- Computation

Suppose n data values are randomly selected from a population with a symmetric distribution. As in the case of the sign test, we consider three scenarios – a left-sided alternative, a right-sided alternative, or a two-sided alternative.

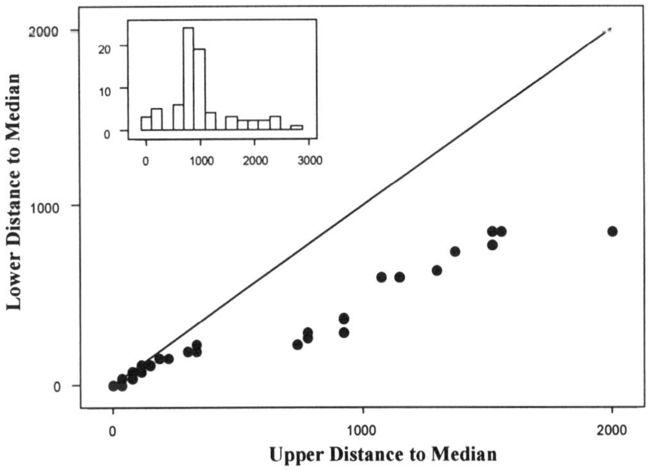

Figure 6.7. Symmetry plot for Tritium data

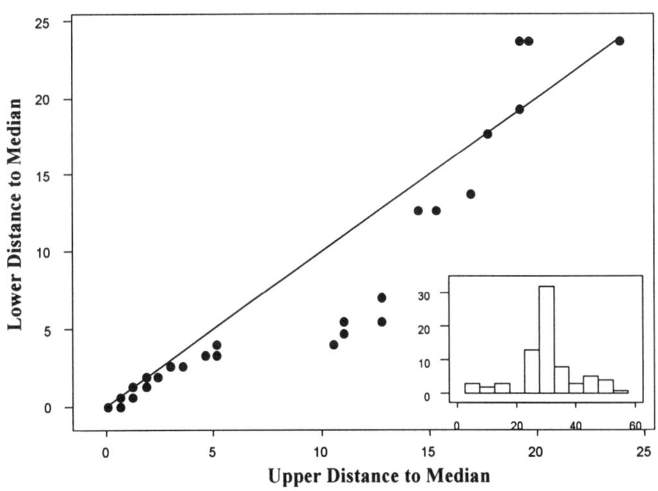

Figure 6.8. Symmetry plot for Tritium data (after square root transformation)

6.5.2.1 ***Left-sided Alternative.*** We consider testing the following pair of hypotheses:
Null Hypothesis: The population median concentration is greater than or equal to x_0

Alternative Hypothesis: The population median concentration is less than x_0

As before, x_0 is a reference value or a regulatory value. Let m denote the number of data values that are different from x_0 (effective sample size). Denote these m data values by $X_1, X_2, \cdots, X_m$. The computation proceeds as follows.

1. Consider the absolute differences $|X_1 - x_0|, |X_2 - x_0|, \cdots, |X_m - x_0|$ of the data values from the reference value. Let $R_1, R_2, \cdots, R_m$ denote their ranks. The data value for which the absolute difference is the smallest is given the rank 1, the next smallest absolute difference is given the rank 2, and so on. Thus, R_1 is the rank given to $|X_1 - x_0|$, R_2 is the rank given to $|X_2 - x_0|$, etc.
2. First consider the case when there are no *ties* among the absolute differences, i.e., the values $|X_1 - x_0|, |X_2 - x_0|, \cdots, |X_m - x_0|$ are all distinct. In this case, compute the test statistic

$$W^+ = \text{Sum of the } R_i \text{ values corresponding to positive differences } X_i - x_0$$

Small values of W^+ indicate that the positive differences are few or small, which supports the alternative hypothesis. Large values of W^+ would be consistent with the null hypothesis. Therefore the null hypothesis is rejected if W^+ is strictly smaller than a critical value, which we denote by w_α. The critical value will depend both on the effective sample size m and the type-I error rate α. Table A12 in Conover (1999) gives these critical values for selected values of α, for (effective) sample sizes between 4 and 50.

3. If there are ties among the values $|X_1 - x_0|, |X_2 - x_0|, \cdots, |X_m - x_0|$, then *average ranks* are assigned to the "tied" values. In this case, critical values given in Table A12 in Conover (1999) will lead to approximate tests. The approximation is satisfactory unless there are too many ties. It is recommended that a suitable computer program (MINITAB) be used when there are "ties" or when the sample size is larger than 50.

6.5.2.2 Right-sided Alternative. Here we consider testing the following pair of hypotheses:

Null Hypothesis: The population median concentration is less than or equal to x_0

Alternative Hypothesis: The population median concentration is greater than x_0

Step 1 in the computation is the same as in the case of a left-sided alternative. In step 2, instead of W^+, we calculate W^- = Sum of the R_i values corresponding to negative differences $X_i - x_0$. Small values of W^- indicate that the negative differences are few or small, which supports the alternative hypothesis. Therefore the null hypothesis is rejected if W^- is strictly smaller than the critical value w_α which can obtained from Table A12 of Conover (1999) corresponding to the selected value of α and (effective) sample size m. Ties are handled as in step 3 for a left-sided alternative.

6.5.2.3 Two-sided Alternative. Here we consider testing the following pair of hypotheses:

Null Hypothesis: The population median concentration is equal to x_0

Alternative Hypothesis: The population median concentration is not equal to x_0

Step 1 in the computation is the same as for a left-sided alternative. In step 2, we calculate both W^+ and W^-. In this case, both small values of W^+ and small values of W^- support the alternative hypothesis. Therefore the null hypothesis is rejected if either W^+ is smaller than a critical value $w_{\alpha/2}$ or W^- is smaller than $w_{\alpha/2}$, where $w_{\alpha/2}$ is obtained from Table A12 in Conover (1999) for (effective) sample size m. Ties in the ranks are handled as in step 3 for a left-sided alternative.

The reason for using $\alpha/2$ when determining the critical value for the two-sided test is that, under a true null hypothesis, the probability will be (approximately) $\alpha/2$ that $w_{\alpha/2}$ will exceed W^+ and (approximately) $\alpha/2$ that $w_{\alpha/2}$ will exceed W^-. This will guarantee that the overall type-I error rate is (approximately) equal to the desired value α. Because the Wilcoxon test statistic can take only integer values, only certain values of α are realizable. Therefore, the desired type-I error rates are only approximately achieved.

We use an artificial data set to illustrate the computations.

6.5.3 Illustrative Example – Wilcoxon Signed Ranks Test

Drinking water supply system in a city is routinely tested for pollutant levels to ensure that it is safe for human consumption. Suppose water samples are obtained and ^{90}Sr concentrations are measured. Suppose the maximum contamination limit (MCL) for ^{90}Sr in drinking water is 296 Bq/m^3. Eleven test tube samples of water are randomly selected from the supply system and ^{90}Sr concentrations measured. The sample data (in Bq/m^3) are given below.

473.6 236.8 384.8 399.6 403.3 488.4 236.8 296.0
222.0 384.8 414.4

We wish to test the following pair of hypotheses.

Null Hypothesis: The median concentration is less than or equal to 296 Bq/m^3
Alternative Hypothesis: The median concentration is greater than 296 Bq/m^3

Here we have $x_0 = 296$. Since the alternative is right-sided, we calculate the test statistic W^-. Table 6.1 lists the intermediate quantities needed to calculate W^-. Notice that there is one data value that is exactly equal to 296, resulting in a difference $X_i - x_0$ equal to 0. This zero value is removed prior to assigning ranks to the absolute differences. Therefore $n = 11$ and $m = 10$ in this problem.

The value of the test statistic W^- is the sum of the ranks corresponding to the negative differences in the column labeled $X_i - x_0$. The ranks that are summed are shown in boldface in Table 6.1. The value of the test statistic is 1.5+1.5+3 = 6. Observe how the "tied" ranks are calculated. For instance, there are two absolute differences each equal to 1.6. The rank positions corresponding to these two 1.6's are 1 and 2 whose average is 1.5. Hence both of these observations are assigned a rank of 1.5. Similar calculations apply to other tied ranks.

Table 6.2. Details of the calculation of Wilcoxon signed-ranks test statistic for the ^{90}Sr data (Bq/m^3).

X_i	$X_i - x_0$	$X_i - x_0$ **with 0 removed**	$\lvert X_i - x_0 \rvert$	**Ranks**	Sign of $X_i - x_0$
473.6	177.6	177.6	177.6	9.0	+
236.8	-59.2	-59.2	59.2	**1.5**	-
384.8	88.8	88.8	88.8	4.5	+
399.6	103.6	103.6	103.6	6.0	+
403.3	107.3	107.3	107.3	7.0	+
488.4	192.4	192.4	192.4	10.0	+
236.8	59.2	-59.2	59.2	**1.5**	-
296.0	0.0	*	*	*	*
222.0	-74.0	-74.0	74.0	**3.0**	-
384.8	88.8	88.8	88.8	4.5	+
414.4	118.4	118.4	118.4	8.0	+

The rejection rule for the hypothesis test is to reject the null hypothesis if W^- is less than the critical value $w_\alpha = w_{0.05} = 11$, obtained from Table A12 in Conover (1999). Since $T^- = 6$ is less than 11, we reject the null hypothesis and conclude that the true median concentration of ^{90}Sr in the water supply is greater than 296 Bq/m^3. Note that this is an approximate test because of the ties. MINITAB may be used to carry out the calculations for the Wilcoxon signed ranks test. Results obtained using MINITAB for the strontium example are shown in Exhibit 6.3.

Exhibit 6.3. Wilcoxn Signed-Ranks test for the ^{90}Sr data

Wilcoxon Signed Rank Test

Test of median = 296.0 versus median > 296.0

	N	N for Test	Wilcoxon Statistic	P	Estimated Median
Strontium	11	10	49.0	0.016	355.2

Minitab reports a P-value equal to 0.016. Since this P-value is smaller than the chosen value for α, the null hypothesis is rejected. Observe that MINITAB reports the value of W^+, which is equal to 49, whereas we calculated $W^- = 6.0$.

In many applications it is desirable to have an interval estimate of the population median. Just as it was possible to get a confidence interval based on the sign test, it is also possible to obtain a confidence interval for the median that is derived from the Wilcoxon signed ranks test. Most statistical packages have a built in command to obtain this. Exhibit 6.4 displays the confidence interval obtained using MINITAB for the problem under discussion.

Exhibit 6.4. Signed-ranks confidence interval for the median using the ^{90}Sr data

Wilcoxon Signed Rank Confidence Interval

	N	Estimated Median	Achieved Confidence	Confidence Interval
Strontium	11	355	95.5	(303, 437)

Observe that the confidence interval for the population median does not include the value 296. The reported confidence interval in this problem has a coverage probability equal to 95.5% even though a 95% interval was requested. Due to the discrete nature of the test statistic, not all confidence levels are achievable.

6.6 INFERENCE FOR PROPORTIONS

There are a number of practical problems where it is of interest to test hypotheses about a population proportion such as the proportion of days when the concentration of a pollutant exceeds a reference value, the proportion of individuals who are exposed to unacceptable, elevated levels of Radon, etc. In this subsection we discuss tests for proportions and illustrate the computations using a real data set. Even though the statistical inferences about proportions make use of the well-known parametric family of distributions called binomial distributions, we include them here under nonparametric methods because the underlying distributions of pollutant concentrations are allowed to be rather general. We begin with an example data set.

6.6.1 Thorium Exposure Example

Workers who routinely carry out grinding and welding operations using thoriated electrodes face the risk of unwarranted high exposures to ^{232}Th, say greater than 1 Annual Limit of Intake (ALI). Suppose π denotes the proportion of workers in the population that face exposures exceeding 1 ALI. We may wish to estimate this proportion and test whether or not it is greater than some reference value which, for illustrative purposes, we take as 0.3. The following data give the total exposure assessment index for 26 workers for intake of ^{232}Th, during grinding and welding using thoriated electrodes (T. Ludwig, et. al., Health Phys. 1999).

5.39, 1.76, 0.01, 0.35, 0.03, 0.30, 1.31, 0.30, 0.29, 0.25, 0.03, 0.23, 0.06, 0.23, 0.75, 0.09, 0.21, 0.04, 0.11, 2.51, 10.20, 1.45, 0.11, 0.71, 0.23, 0.53

For this example, the exposure assessment index is defined as the ratio of the determined annual intake divided by ALI. Suppose these 26 workers are a random sample from the population of all such workers. Six of the 26 workers in the sample have assessment indices that exceed 1. We wish to use the sample data to test the following pair of hypotheses.

Null Hypothesis: π is greater than or equal to 0.3

Alternative Hypothesis: π is less than 0.3

The estimated proportion of workers in the population with assessment index exceeding 1 is 6/26 = 0.231, which lends some support to the alternative hypothesis that π is less than 0.3. However, is the evidence provided by the data strong enough for us to rule out the possibility that π is greater than or equal to 0.3? We proceed to discuss methods for answering such questions involving proportions.

6.6.2 Estimates and Hypothesis Tests for Proportions

Let $X_1, X_2, \cdots, X_n$ represent the data values, which are assumed to be a random sample from the population of interest. Suppose π is the proportion of values in the population that satisfy a requirement. For instance, π might refer to the proportion of values that exceeds a reference value x_0, or the proportion that is less than a reference value x_0, or the proportion that lies between two reference values x_0 and x_1, etc. The estimate of π is given by the corresponding sample proportion s/n where s is the number of sample values that satisfy the requirement and n is the number of observations in the sample. As usual, the alternative hypothesis for the test may be left-sided, right-sided, or two-sided.

6.6.2.1 Left-sided Alternative. In this case we are interested in testing
Null Hypothesis: π is greater than or equal to π_0

Alternative Hypothesis: π is less than π_0

Let T^+ denote the number of sample data values that satisfy the requirement of interest. Observe that small values of T^+ support the alternative hypothesis. Hence the null hypothesis is rejected if the value of T^+ is smaller than a critical value, which we denote by T^*. The value T^* depends on the reference proportion π_0, the sample size n, and the value of α used for the type I error rate. It is impractical to tabulate the critical values for every possible situation. However an approximate P-value for this test may be computed by first calculating

$$Z_0 = \frac{T^+ + 0.5 - n\pi_0}{\sqrt{n\pi_0(1-\pi_0)}} \tag{6.1}$$

and then determining the area to the left of Z_0 under the standard normal curve. If the P-value is smaller than the chosen value for α then the null hypothesis is rejected. Areas for the standard normal curve are tabulated in nearly every statistics book. They may also be obtained using standard software packages. Alternatively, (nearly) exact P-values for the test may be obtained from MINITAB and other leading statistical packages.

6.6.2.2 Continuity Correction. The term 0.5 in the numerator of the expression for Z_0 in Eq. (6.1) is called a *continuity correction* term. Note that T^+ can only take integer values. However, we are approximating its distribution using a normal distribution, which is a continuous distribution. In this approximation, an integer t in the discrete distribution corresponds to the entire interval $t-0.5$ to $t+0.5$ in the approximating normal distribution. So, the probability that the value of T^+ will be less than or equal to an integer t is approximated well by considering the area to the left of $t + 0.5$ in the approximating normal distribution, rather than just the area to the left of t. If we need the probability that T^+ will take a value greater than or equal to an integer t, then we use the area to the right of $t - 0.5$ in the approximating normal distribution, rather than just the area to the right of t.

6.6.2.3 ***Right-sided Alternative.*** In this case we are interested in testing

Null Hypothesis: π is less than or equal to π_0

Alternative Hypothesis: π is greater than π_0

Let T^- denote the number of sample data values that do not satisfy the criterion of interest. Again, small values of T^- support the alternative hypothesis. As before, an approximate P-value for this test may be computed by first calculating

$$Z_0 = \frac{T^- + 0.5 - n\pi_0}{\sqrt{n\pi_0(1-\pi_0)}} \tag{6.2}$$

and then determining the area to the left of Z_0 under the standard normal curve. Alternatively, (nearly) exact P-values for the test may be obtained from MINITAB or other leading statistical packages.

6.6.2.4 ***Two-sided Alternative.*** In this case we are interested in testing

Null Hypothesis: π is equal to π_0

Alternative Hypothesis: π is not equal to π_0

Let T^+ denote the number of sample data values that satisfy the criterion of interest and T^- denote the number of sample data values that do not satisfy the criterion of interest. Small values of T^+ as well as small values of T^- support the alternative hypothesis. An approximate P-value for this test may be computed as follows. Calculate

$$Z_1 = \frac{T^+ + 0.5 - n\pi_0}{\sqrt{n\pi_0(1-\pi_0)}} \tag{6.3}$$

and

$$Z_2 = \frac{T^- + 0.5 - n\pi_0}{\sqrt{n\pi_0(1-\pi_0)}} \tag{6.4}$$

and determine the area A_1 to the left of Z_1 and the area A_2 to the left of Z_2 under the standard normal curve. Then the P-value for this test is (approximately) given by 2 times the smaller of A_1 and A_2. Alternatively, (nearly) exact P-values for the test may be obtained from MINITAB or other leading statistical packages.

We now use the thorium exposure example to illustrate the calculations.

6.6.3 Illustrative Example: Thorium Data

For the thorium data set, we are interested in a left-sided alternative. The value of the test statistic T^+ is 6. Hence $Z_0 = \frac{6+0.5-(26)(0.3)}{\sqrt{(26)(0.3)(1-0.3)}}$ = -0.5563. From a table of areas under the standard normal curve we determine the P-value to be 0.289. If we choose α to be 0.05, then we cannot reject the null hypothesis because the P-value is not less than the chosen value of α. Thus, we cannot rule out the possibility that the true proportion of workers with assessment indices above 1 is greater than or equal to 0.3. Exhibit 6.5 displays the MINITAB output for this test.

Exhibit 6.5. Test of a proportion in the thorium exposure example

Test and Confidence Interval for One Proportion

Test of p = 0.3 vs p < 0.3

Sample	X	N	Sample p	95.0 % CI	Exact P-Value
1	6	26	0.230769	(0.089740, 0.436475)	0.297

The exact P-value is 0.297 which is quite close to the approximate value computed in this example. The approximation will improve as the sample size n becomes larger.

Note that MINITAB also displays a two-sided 95% confidence interval for the proportion of interest. We are thus 95% confident that the true proportion of individuals whose exposure to thorium exceeds 1 ALI lies between about 9% and 44%. Therefore we cannot rule out the possibility that this proportion exceeds 0.3.

6.7 MANN-WHITNEY TEST

This test is also called the Wilcoxon Rank Sum test and is used for testing whether or not one population median (or mean) is greater than the population median (or mean) of another population. A typical application of this test is for comparing the mean or median value of a variable of interest at a test site with the corresponding value at a reference site.

6.7.1 Computation for Mann-Whitney Test

A random sample of size n is selected from the first population and, independently, a random sample of size m is selected from the second population. Let $X_1, X_2, \cdots, X_n$ denote the random sample from population 1 and $Y_1, Y_2, \cdots, Y_m$ denote the random sample from population 2. Note that the sample sizes for the two samples may or may not be equal. As usual, the alternative hypothesis may take one of the three forms – a left-sided alternative, a right-sided alternative, or a two-sided alternative. It is assumed that the two populations being compared have the same distribution, except possibly for a difference in their locations. This assumption implies that the difference between the medians of the two populations is equal to the difference between their means. Let Median_1 denote the median of population 1 and Median_2 denote the median of population 2.

6.7.1.1 Left-sided Alternative. In this case the hypotheses being tested are
Null Hypothesis: Median_1 is greater than or equal to Median_2.
Alternative Hypothesis: Median_1 is less than Median_2.

The test is carried out as follows.

1. Combine the two sample data sets to form one single data set consisting of $n+m$ observations. Assign the ranks 1 to $n+m$ to these data values, rank 1 being assigned to the smallest value and rank $n+m$ assigned to the largest value. (If several sample values are equal to one another, then assign each of the tied values the *average of the ranks* that would be assigned had there been no ties.)
2. Calculate the sum of the ranks assigned to the sample data from population 1 and denote this sum by W_1.
3. Small values of W_1 support the alternative hypothesis. Hence the null hypothesis will be rejected if the calculated value of W_1 is less than a critical value which we denote by W^*. This value depends on the sample sizes n and m as well as the value of α. Table A7 of Conover (1999) gives these critical values for selected values of α, for n and m less than or equal to 20. The critical value is the entry in the row labeled n and column labeled m corresponding to the chosen value of α.

6.7.1.2 Right-sided Alternative. In this case the hypotheses being tested are
Null Hypothesis: Median_1 is less than or equal to Median_2.
Alternative Hypothesis: Median_1 is greater than Median_2.
The test is carried out as follows.

1. This step is the same as step 1 for the left-sided alternative.
2. Let W_1 denote the sum of the ranks assigned to sample values from population 1. In this case large values of W_1 support the alternative hypothesis and so the critical values needed are different from those in the case of a left-sided alternative. Rather than give a new set of critical values in a separate table, we change the test statistic. We use the statistic W_2 which is defined as the sum of the ranks assigned to the sample data from population 2. It is useful to note that W_1+W_2 equals $(n+m)(n+m+1)/2$.
3. Small values of W_2 support the alternative hypothesis. Hence the null hypothesis will be rejected if the calculated value of W_2 is less than the critical value W^*

which may be obtained from Table A7 of Conover (1999) by looking up the entry in the row labeled m and column labeled n (i.e., the roles of m and n are interchanged) for the chosen value of α.

6.7.1.3 Two-sided Alternative. In this case the hypotheses being tested are
Null Hypothesis: Median_1 is equal to Median_2.
Alternative Hypothesis: Median_1 is not equal to Median_2.

The test is carried out as follows.

1. Rank the combined data as in step 1 for the left-sided alternative.
2. Let W_1 denote the sum of the ranks assigned to the sample data from population 1 and W_2 denote the sum of the ranks assigned to the sample data from population 2.
3. Small values of W_1 and small values of W_2, both support the alternative hypothesis. The null hypothesis is rejected if either W_1 is less than the critical value W_1^*, or if W_2 is less than the critical value W_2^*. In this case, the critical value W_1^* is the entry for $\alpha/2$ in the cell corresponding row n and column m and the critical value W_2^* is the entry for $\alpha/2$ in the cell corresponding row m and column n (i.e., the roles of m and n are interchanged).

Tests using the critical values given in Table A7 of Conover (1999) are exact only when there are no ties. If there are too many ties, or if the sample sizes are large, it is recommended that a suitable computer package be used to carry out this test. MINITAB has a built in command for this test, as do most of the leading statistical packages.

We illustrate the computations for the Mann-Whitney test using the data derived from the article titled "Impact of a 2 MW_{th} Research Reactor on Radioactivity in Sediments" by S.R. Bose, et. al. (Health Phys. 1993).

6.7.2 Reactor Data Example

A study was conducted to determine the environmental impact caused by liquid effluent discharge from the University of Virginia's 2 MW_{th} research reactor during the course of its first 25 years of normal operation. Sediment samples were collected from the reactor pond, the stream feeding it and its exit stream. For purposes of comparison, sediment samples were also taken from a nearby reference pond having no direct link with the reactor pond. Table 6.3 displays gross alpha radioactivity (Bq/kg dry sediment) in the sediments. The first column gives the measurement for a single sample collected upstream from the reactor pond; the second column gives the measurements for the 6 samples collected in the reactor pond; the third column gives the 5 readings obtained from the exit pond; and the last column gives the measurements for the 2 sediment samples obtained from the reference pond.

Table 6.3. Gross alpha radioactivity (Bq/kg dry sediment) in the sediment samples collected at four different sites around a research reactor.

Upstream	Reactor Pond	Exit Stream	Reference Pond
966	1400, 953, 1577, 1430, 1710, 1250	994, 938, 987, 636, 484	479, 468

For the purpose of illustration we will compare the Reactor Pond (test site) measurements with the reference pond measurements. Suppose we wish to test

Null Hypothesis: The median value at the test site is less than or equal to that at the reference site
Alternative Hypothesis: The median value at the test site is greater than that at the reference site

We have $m = 6$ and $n = 2$ respectively. Combining the two data sets yields a total of 8 observations. The alternative hypothesis is *right-sided* so the test statistic is W_2. Summing the ranks corresponding to the reference site we get $W_2 = 3$. For $\alpha = 0.05$, Table A7 from Conover (1999) yields the critical value 5 from the row labeled 2 and column labeled 6. Since the calculated value of the test statistic is 3, which is less than the critical value, the null hypothesis is rejected. We can therefore conclude that the median gross alpha radioactivity at the test site is greater than that at the reference site.

Exhibit 6.6. Mann-Whitney test for the reactor data.

Mann-Whitney Confidence Interval and Test

Test site N = 6 Median = 1415.0
Reference site N = 2 Median = 473.5
Point estimate for ETA1-ETA2 is 941.5
93.3 Percent CI for ETA1-ETA2 is (473.9,1242.1)
W = 33.0
Test of ETA1 = ETA2 vs ETA1 > ETA2 is significant at 0.0334

Exhibit 6.6 displays the MINITAB output for this hypothesis test. Minitab always reports the sum of the ranks for the sample that is labeled the first sample. Here the sum of the ranks for the first sample is 33 which is the value reported by MINITAB under the name W. Observe also that MINITAB reports a confidence interval for the difference between the two medians (median for population 1 (test site) minus median for population 2 (reference site)). For the present example we can be about 95% confident (93.3% to be exact) that $\text{Median}_{\text{test site}} - \text{Median}_{\text{reference site}}$ is between 473.9 and 1242.1 Bq/kg dry sediment.

6.8 COMPARING TWO MEDIANS BASED ON PAIRED SAMPLES: WILCOXON'S SIGNED-RANKS TEST REVISITED

In the previous section we studied a nonparametric test, namely the Mann-Whitney test, for comparing two population medians based on **two independent samples**. That test is the analog of the parametric two-independent samples t-test. When the two samples are not independent, but each sample from one population (site) is matched in a natural manner with a sample from the second population, it is appropriate to consider the differences between the matched-pairs and make inferences based on these differences. The standard parametric technique for such problems is the matched-pairs t-test. The nonparametric counterpart of this test is based on the Wilcoxon's signed-ranks test discussed earlier.

6.8.1 Computation of Wilcoxon Signed Ranks Test for Paired Data

Let $(X_1,Y_1),(X_2,Y_2),\cdots,(X_n,Y_n)$ denote a random sample of n matched-pairs with the first member of the pair coming from population 1 and the second member coming from population 2. Let D_i denote the difference $X_i - Y_i$. In environmental applications, comparisons are often made between a test site (population 1) and a reference site (population 2). For instance, pairs of samples may be collected on randomly selected days within a year with one sample from the test site and the other from the reference site. The two samples are not independent in this situation because the same sources of variability might influence the same day's measurement from both sites. In this case it is appropriate to examine the differences $D_i = X_i - Y_i$ (perhaps after transforming the data suitably) and ask the question whether the median difference is zero, less than zero, or greater than zero. The Signed-Ranks test for paired-data is based on the assumption that the population of differences D_i is symmetric. Let Median_1 denote the median of population 1 and Median_2 denote the median of population 2.

6.8.1.1 Left-sided Alternative. The hypotheses considered in this case are
Null Hypothesis: Median_1 is greater than or equal to Median_2.
Alternative Hypothesis: Median_1 is less than Median_2.

6.8.1.2 Right-sided Alternative. The hypotheses considered in this case are
Null Hypothesis: Median_1 is less than or equal to Median_2.
Alternative Hypothesis: Median_1 is greater than Median_2.

6.8.1.3 Two-sided Alternative. The hypotheses considered in this case are
Null Hypothesis: Median_1 is equal to Median_2.
Alternative Hypothesis: Median_1 is not equal to Median_2.

In each case, the test is carried out in a fashion very similar to that in Sec. 6.5 using the differences $D_i = X_i - Y_i$. Let m denote the number of nonzero differences. We will refer to m as the effective sample size. We use an artificial data set to illustrate the computations.

6.8.2 Illustrative Example of Wilcoxon's Signed Ranks Test for Paired Data

A study was conducted to test the research hypothesis that Radon concentrations in the basements of dwellings in high Radon risk areas are generally higher than concentrations on the main floor. Let $\text{Median}_{\text{basement}}$ and $\text{Median}_{\text{main floor}}$ denote the median Radon concentrations for basements and main floors, respectively. The following pair of hypotheses was considered.

Null Hypothesis: $\text{Median}_{\text{basement}}$ is less than or equal to $\text{Median}_{\text{main floor}}$.
Alternative Hypothesis: $\text{Median}_{\text{basement}}$ is greater than $\text{Median}_{\text{main floor}}$.

Seventeen dwellings with basements were randomly selected from a particular high-risk area and two Radon measurements (in Bq/m^3) were made in each dwelling – one in the basement and one on the main floor. The data, along with calculations needed to obtain the test statistic, are given in Table 6.4.

In this case, we calculate T^- (the sum of the ranks corresponding to the negative differences) as the test statistic. Small values of T^- support the alternative hypothesis. The sum of the ranks corresponding to the negative differences, shown boldfaced in Table 6.4, is 6.5. The critical value from Table A12 of Conover (1999), using effective sample size $m = 15$ and $\alpha = 0.05$, is 31. Since 6.5 is smaller than the critical value the null hypothesis is rejected. The data provide evidence supporting the alternative hypothesis. MINITAB can be used to carry out these computations. The output from MINITAB is shown in Exhibit 6.7.

Exhibit 6.7. Calculation of Wilcoxon's signed-ranks test for the Basement-main floor radon data

Wilcoxon Signed Rank Test

Test of median = 0.000000 versus median > 0.000000

	N	N for Test	Wilcoxon Statistic	P	Estimated Median
Difference	17	15	113.5	0.001	166.5

Table 6.4. Details of the calculation of Wilcoxon signed-ranks test statistic for the radon data (Bq/m^3).

Dwelling	**Basement**	**Main Floor**	**Difference: Basement minus Main Floor**	**Difference (0 omitted)**	**Absolute difference**	**Rank of absolute difference**	**Sign of difference**
1	518	37	481	481	481	15.0	+
2	518	296	222	222	222	9.0	+
3	407	111	296	296	296	13.0	+
4	444	0	444	444	444	14.0	+
5	259	185	74	74	74	2.5	+
6	222	222	0	*	*	*	*
7	296	74	222	222	222	9.0	+
8	370	555	-185	-185	185	**5.5**	-
9	444	259	185	185	185	5.5	+
10	518	444	74	74	74	2.5	+
11	370	407	-37	-37	37	**1.0**	-
12	481	259	222	222	222	9.0	+
13	259	37	222	222	222	9.0	+
14	518	259	259	259	259	12.0	+
15	407	296	111	111	111	4.0	+
16	370	148	222	222	222	9.0	+
17	444	444	0	*	*	*	*

Observe that MINITAB reports the sum of the ranks corresponding to the positive differences as the test statistic. This is equal to 113.5. The P-value is 0.001. Since the value of α chosen for the test is 0.05 and the P-value is less than 0.05, the null hypothesis is rejected. A two-sided 95% confidence interval for the true difference between the two medians (basement minus main floor) is given in Exhibit. 6.8.

Exhibit 6.8. Wilcoxon's signed-ranks confidence interval for the median difference between basements and main floors in the radon study.

Wilcoxon Signed Rank Confidence Interval

	N	Estimated Median	Achieved Confidence	Confidence Interval
Difference	17	167	94.8	(93, 241)

We can be about 95% confident (actually 94.8%) that the median Radon concentration for basements of dwellings is between 93 Bq/m^3 to 241 Bq/m^3 higher than the median Radon concentration for main floors.

6.9 KRUSKAL-WALLIS K-SAMPLE TEST

When the median values of more than two populations are to be compared, an extension of the Mann-Whitney test is available, and it is called the Kruskal-Wallis test for comparing k populations based on k independent samples. It tests the following pair of hypotheses.

Null Hypothesis: The medians of the k populations are all equal to one another.
Alternative Hypothesis: At least one population median is different from the rest.

6.9.1 Computation of the Kruskal-Wallis Test

Let $X_{1,1}, X_{1,2}, \cdots, X_{1,n_1}$ denote the sample data from population 1, $X_{2,1}, X_{2,2}, \cdots, X_{2,n_2}$ denote the sample data from population 2, etc., and $X_{k,1}, X_{k,2}, \cdots, X_{k,n_k}$ denote the sample data from population k. The sample sizes for the k groups need not be equal. These sample sizes are $n_1, n_2, \cdots, n_k$, respectively.
The test procedure is as follows.

1. Let $N = n_1 + n_2 + \cdots + n_k$ denote the total sample size.
2. Combine all N data values, order them, and rank them. The smallest data value is assigned the rank 1, the next smallest is assigned the rank 2, and so on. If there are ties, then the average rank is assigned to each of the tied values.

3. Let R_i denote the sum of the ranks assigned to the observations in the ith sample.
4. The test statistic is defined by

$$H = \frac{1}{S^2}\left(\sum_{i=1}^{k} \frac{R_i^2}{n_i} - \frac{N(N+1)^2}{4} \right) \tag{6.5}$$

where

$$S^2 = \text{ the sample variance of all the } N \text{ ranks.} \tag{6.6}$$

If there are no ties then it can be shown that

$$S^2 = \frac{N(N+1)}{12}. \tag{6.7}$$

An alternative expression for the test statistic is given by

$$H = \frac{\sum n_j (\overline{R}_j - \overline{R})^2}{S^2} \tag{6.8}$$

where $\overline{R}_j$ is the average rank for sample j and $\overline{R}$ is the average rank for all sample data combined. It follows from this alternative form of H that it is always greater than or equal to zero. It is equal to zero precisely when the average ranks for all the groups are equal. Thus, small values of H are consistent with the null hypothesis, whereas large values of H support the alternative hypothesis. Hence, the null hypothesis will be rejected when H is greater than or equal to a critical value, which we denote by H^*. These critical values are too numerous to tabulate conveniently. However, they may be computed approximately by referring to a χ^2 distribution table with $k-1$ degrees of freedom and the chosen value for α.

When there are ties, some authors suggest the use of an adjustment for the statistic H. This adjusted H is given by

$$H_{adj} = \frac{H}{1 - \frac{\sum_{j=1}^{R}(d_j^3 - d_j)}{N^3 - N}} \tag{6.9}$$

where R is the number of *distinct* data values and d_j is the number of tied data values in the jth group of tied data values, $j = 1,2,\cdots,R$. The χ^2 distribution with $k-1$ degrees of freedom may be used in this case as well to determine (approximate) critical values. Most leading statistical packages have a built in command for computing the Kruskal-Wallis test, as does MINITAB. As a rule, these computer programs will make the necessary adjustment for ties.

When the null hypothesis is rejected, it is generally of interest to determine which pairs of population medians are different. A commonly used (approximate) procedure is to decide that population medians for population i and population j are different whenever the following inequality holds:

$$\left|\frac{R_i}{n_i} - \frac{R_j}{n_j}\right| > t_{1-\frac{\alpha}{2},N-k}\left(\frac{S^2(N-1-H)}{N-k}\right)^{1/2}\left(\frac{1}{n_i} + \frac{1}{n_j}\right)^{1/2} \tag{6.10}$$

where $t_{1-\frac{\alpha}{2}}$ refers to the t-table value with k-1 degrees of freedom such that the area to the right of it is equal to $\alpha/2$. The left side of Eqn. (6.10) may be regarded as the test statistic for comparing populations i and j, and the right side of Eqn. (6.10) as the critical value. This procedure is precisely Fisher's F-protected LSD test applied to the ranks of the data. Thus, the calculations can be carried out using any computer program which can perform multiple-comparisons of groups.

6.9.2 Thorium Study Revisited

In the thorium study presented earlier, the 26 workers could be divided into four groups based on the type of current used (AC or DC) and the type of electrode used

(WT20 or WT40). The WT40 type electrode contained twice the mass fraction of Thorium as the WT20 type. The data for the 26 workers is given in Table 6.5.

We test the null hypothesis that the true median assessment index for the four groups are equal versus the alternative hypothesis that at least one median is different from the rest. The sample median assessment indices for the four groups are -- group 1: 0.3750; group 2: 0.9900; group 3: 0.2300; and group 4: 0.2900. Observe that group 2 (DC current; electrode type WT40) appears to have a substantially higher median assessment index than the other three groups. We now examine the statistical significance of this observation using the Kruskal-Wallis test.

The sums of the ranks for the four groups are $R_1 = 50.5$, $R_2 = 108.5$, $R_3 = 125.0$, $R_4 = 67.0$, respectively. The sample variance of the ranks is $S^2 = 58.36$. Hence the test statistic is $H = 3.07332$. The critical value corresponding to $\alpha = 0.05$ and degrees of freedom = 4 – 1 = 3 is obtained from a Chi-square table. It is equal to 7.815. Therefore the data do not provide sufficient evidence to establish differences between the medians of the four groups! Due to the lack of significance, we do not follow up this analysis by pairwise comparison of individual groups.

Exhibit 6.9 shows the results of the Kruskal-Wallis test obtained using MINITAB. The test statistic is reported as $H = 3.07$ (compared to 3.07332 calculated earlier) and the P-value is equal to 0.380. Since the P-value is not smaller than 0.05 used for the value of alpha, the null hypothesis cannot be rejected.

Exhibit 6.9. MINITAB output for Kruskal-Wallis test using thorium data

Kruskal-Wallis Test

Kruskal-Wallis Test on Assessment Index

Group	N	Median	Ave Rank	Z
1	4	0.3750	12.6	-0.25
2	6	0.9900	18.1	1.67
3	11	0.2300	11.4	-1.22
4	5	0.2900	13.4	-0.03
Overall	26		13.5	

H = 3.07 DF = 3 P = 0.382
H = 3.07 DF = 3 P = 0.380 (adjusted for ties)
* NOTE * One or more small samples

Table 6.5. Thorium exposure data

Worker	AC=1 DC=2	WT40=1 WT20=2	Group	Assessment Index Welding+Grinding	Ranks
1	1	1	1	5.39	25.0
2	1	1	1	0.03	2.5
3	1	1	1	0.04	4.0
4	1	1	1	0.71	19.0
5	1	2	2	2.51	24.0
6	1	2	2	10.2	26.0
7	1	2	2	1.45	22.0
8	1	2	2	0.11	7.5
9	1	2	2	0.23	11.0
10	1	2	2	0.53	18.0
11	2	1	3	1.76	23.0
12	2	1	3	0.01	1.0
13	2	1	3	0.30	15.5
14	2	1	3	1.31	21.0
15	2	1	3	0.03	2.5
16	2	1	3	0.23	11.0
17	2	1	3	0.06	5.0
18	2	1	3	0.23	11.0
19	2	1	3	0.75	20.0
20	2	1	3	0.09	6.0
21	2	1	3	0.21	9.0
22	2	2	4	0.35	17.0
23	2	2	4	0.30	15.5
24	2	2	4	0.29	14.0
25	2	2	4	0.25	13.0
26	2	2	4	0.11	7.5

6.9.3 Reactor Pond Example Revisited

We now revisit the University of Virginia Reactor Pond data discussed earlier and compare the population medians for the three sites – reactor pond, exit pond, and reference pond. Because there was only one data value for the upstream pond we do not consider this as a fourth group. The combined ranks and the sum of the ranks for each group are given below. The Site ID values are defined as follows:

1 = reactor pond, 2 = exit stream, 3 = reference pond.

The response variable is gross alpha concentrations (Bq/kg dry sediment).

Table 6.6. Gross alpha measurements (Bq/kg dry sediment) for three different sites.

Site ID	Gross Alpha	Rank	Site ID	Gross Alpha	Rank
1	1400	10	2	994	8
1	953	6	2	938	5
1	1577	12	2	987	7
1	1430	11	2	636	4
1	1710	13	2	484	3
1	1250	9	3	479	2
			3	468	1

The sums of the ranks for the three groups are $R_1 = 61, R_2 = 27, R_3 = 3$, respectively. Since there are no ties, the value of S^2 is equal to $\frac{(13)(14)}{12} = 15.167$. Hence the value of the test statistic is

$$H = \frac{1}{15.167}\left(\frac{61^2}{6} + \frac{27^2}{5} + \frac{3^2}{2} - \frac{(13)(14)^2}{4}\right) = 8.799\,.$$

The critical value corresponding to $\alpha = 0.05$, and degrees of freedom equal to (3-1) = 2 is obtained from a χ^2 distribution table, and is equal to $H^* = 5.991$. Since the value of H is greater than H^*, the null hypothesis is rejected and we conclude that the true medians for the three locations are in fact different.

Calculations for all pairwise comparisons of the three population medians may be organized as shown in Table 6.7.

Table 6.7. All pairwise comparisons of the median gross alpha activity among the three sites.

Populations i versus j	**Test Statistic** $\left\lvert \frac{R_i}{n_i} - \frac{R_j}{n_j} \right\rvert$	**Critical Value** $t_{1-\frac{\alpha}{2},N-k}\left(\frac{S^2(N-1-H)}{N-k}\right)^{1/2}\left(\frac{1}{n_i}+\frac{1}{n_j}\right)^{1/2}$
1 and 2	$\left\lvert \frac{61}{6} - \frac{27}{5} \right\rvert = 4.767$	$2.228\left(\frac{(15.1667)(13-1-8.799)}{13-3}\right)^{1/2}\left(\frac{1}{6}+\frac{1}{5}\right)^{1/2} = 2.973$
1 and 3	$\left\lvert \frac{61}{6} - \frac{3}{2} \right\rvert = 8.667$	$2.228\left(\frac{(15.1667)(13-1-8.799)}{13-3}\right)^{1/2}\left(\frac{1}{6}+\frac{1}{2}\right)^{1/2} = 4.008$
2 and 3	$\left\lvert \frac{27}{5} - \frac{3}{2} \right\rvert = 3.9$	$2.228\left(\frac{(15.1667)(13-1-8.799)}{13-3}\right)^{1/2}\left(\frac{1}{5}+\frac{1}{2}\right)^{1/2} = 4.107$

From this table we see that the difference between the medians of site 1 and site 2 is statistically significant at $\alpha = 0.05$. The test statistic for comparing site 1 with site 2 is equal to 4.767 which is greater than the corresponding critical value of 2.973. Likewise, the difference between the medians of site 1 and site 3 is statistically significant but not the difference between the medians of sites 2 and 3.

6.10 SPEARMAN CORRELATION

This is a nonparametric measure of association between two variables. Generally, one of the two variables is often thought of as the independent variable and the other as the dependent variable. The Spearman correlation coefficient summarizes the degree to which the two variables are *monotonically* associated. We say that there is a monotonic positive relationship between two variables if it is true that, whenever one variable increases the other shows a tendency to increase as well. There is a monotonic negative relationship between the two if it is true that, whenever one variable increases the other shows a tendency to decrease. Linear relationships are monotonic relationships but monotonic relationships need not be linear. Fig. 6.9 shows a monotonic decreasing relationship and Fig. 6.10 shows a monotonic increasing relationship. Neither relationship is described by a straight line pattern.

6.10.1 Computation of Spearman Correlation Coefficient

Let $(X_1,Y_1),(X_2,Y_2),\cdots,(X_n,Y_n)$ denote n measurement pairs. We wish to investigate whether or not the variable Y is related to the variable X. In parametric statistical analysis, a measure of association that is widely used is the Pearson correlation coefficient between X and Y. This is calculated according to the formula

$$\text{Pearson Correlation Coefficient} = \frac{\sum_{i=1}^{n}(X_i-\overline{X})(Y_i-\overline{Y})}{\left[\sum_{i=1}^{n}(X_i-\overline{X})^2\right]^{1/2}\left[\sum_{i=1}^{n}(Y_i-\overline{Y})^2\right]^{1/2}} \tag{6.11}$$

When the term "correlation" is used (without modifiers) it is usually assumed to refer to the Pearson correlation coefficient. The Pearson correlation coefficient measures the degree of *linear* relationship between X and Y. If this correlation is calculated using ranks of the data rather than the actual values, the resulting number is called the Spearman Rank Correlation Coefficient or Spearman's Rho. This measures the degree of monotonic relationship between X and Y.

Let $R(X_i)$ denote the rank of X_i among all the X values and $R(Y_i)$ denote the rank of Y_i among all the Y values. When there are ties "average ranks" are used in the usual manner. Spearman's Rho for the sample data is given by

$$\text{Rho} = \frac{\sum_{i=1}^{n} R(X_i)R(Y_i) - n\left(\frac{n+1}{2}\right)^2}{\left[\sum_{i=1}^{n} R(X_i)^2 - n\left(\frac{n+1}{2}\right)^2\right]^{1/2}\left[\sum_{i=1}^{n} R(Y_i)^2 - n\left(\frac{n+1}{2}\right)^2\right]^{1/2}} \tag{6.12}$$

Rho can take any value between –1 and +1 (inclusive). Rho is equal to 1 if there is a perfect increasing monotonic relationship or association between X and Y. It is equal to –1 if there is a perfect decreasing monotonic relationship between X and Y. Spearman's Rho for the data shown in Fig. 6.9 is –1 and for the data in Fig. 6.10 it is equal to +1.

A test for association between two variables can be constructed using Spearman's Rho. Statistical independence between X and Y means that neither variable provides useful information for predicting the value of the other. Whenever Y and X are associated, one can generally construct a procedure for predicting Y based on the knowledge of the value of X (or predict X using Y). As usual, three different pairs of hypotheses may be considered.

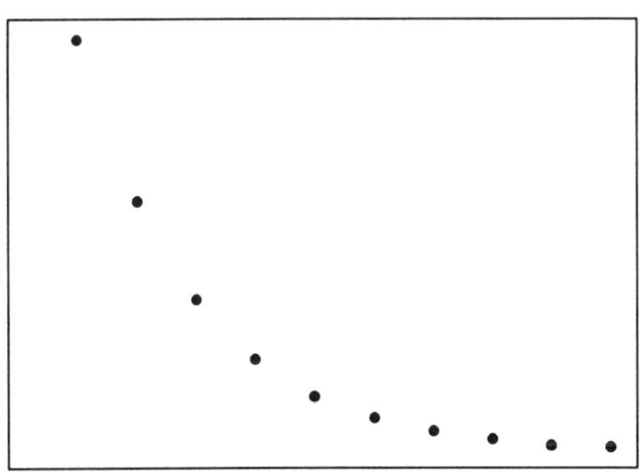

Figure 6.9. A monotonic decreasing relationship

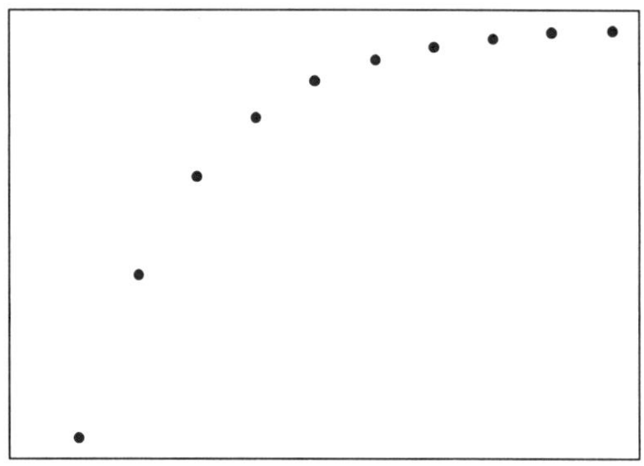

Figure 6.10. A monotonic increasing relationship

6.10.1.1 Left-sided Alternative. In this case the hypotheses are
Null Hypothesis: X and Y are statistically independent of each other.
Alternative Hypothesis: There is a negative monotonic relationship between X and Y.

The test is carried out as follows.

1. Compute Rho as defined above.
2. Values of Rho smaller than zero support the alternative hypothesis. Hence the null hypothesis is rejected if Rho is less than a critical value, which we denote by Rho1*. This critical value depends on the chosen value of α and the sample size n. The value of Rho1* is the negative of the entry for $p = 1 - \alpha$ in Table A10 of Conover (1999) corresponding to the sample size n. This table gives the critical values for sample sizes between 4 and 30 only. For larger sample sizes, the required critical value may be (approximately) calculated as follows.

$$\text{Rho1}^* \approx \frac{z_\alpha}{\sqrt{n-1}} \tag{6.13}$$

where z_α is a value such that the area to the left of it under the standard normal curve is equal to α, i.e., z_α is the αth quantile of the standard normal distribution.

3. The P-value for the test is (approximately) equal to the area to the left of (Rho $\times \sqrt{(n-1)}$) under the standard normal density function.

6.10.1.2 Right-sided Alternative. In this case the hypotheses are
Null Hypothesis: X and Y are statistically independent of each other.
Alternative Hypothesis: There is a positive monotonic relationship between X and Y.

The test is carried out as follows.

1. Compute Rho.
2. Values of Rho larger than zero support the alternative hypothesis. Hence the null hypothesis is rejected if Rho is larger than a critical value, which we denote by Rho2*. This critical value depends on the chosen value of α and the sample size n. The value of Rho2* is the entry for $p = 1-\alpha$ in Table A10 of Conover (1999) corresponding to the sample size n. For sample sizes greater than 30, the required critical value may be (approximately) calculated as follows

$$\text{Rho2}^* \approx \frac{z_{1-\alpha}}{\sqrt{n-1}} \tag{6.14}$$

where $z_{1-\alpha}$ is a value such that the area to the left of it under the standard normal curve is $1-\alpha$.

3. The P-value for the test is (approximately) equal to the area to the right of (Rho $\times \sqrt{(n-1)}$) under the standard normal density function.

6.10.1.3 Two-sided Alternative. In this case the hypotheses are
Null Hypothesis: X and Y are statistically independent of each other.
Alternative Hypothesis: There is a monotonic association between X and Y (this may be a positive or a negative association).

The test is carried out as follows.

1. Compute Rho.
2. Values of Rho different from zero (in either direction) support the alternative hypothesis. Hence the null hypothesis is rejected if Rho is smaller than a critical value, which we denote by Rho1*, or larger than a critical value, denoted by Rho2*. Rho2* is obtained from Table A10 of Conover (1999) by reading the entry for $p = 1-\alpha/2$ corresponding to sample size n, and Rho1* is the negative of Rho2*. For larger sample sizes the required critical value may be (approximately) calculated as follows

$$\text{Rho1}^* \approx \frac{z_{\alpha/2}}{\sqrt{n-1}} \tag{6.15}$$

and

$$\text{Rho2}^* \approx \frac{z_{1-\alpha/2}}{\sqrt{n-1}} \tag{6.16}$$

3. The P-value for the test is **twice** the area to the right (|Rho| × $\sqrt{(n-1)}$) the standard normal density function.

6.10.2 Illustrative Example: Transfer Factors for Radiocesium

Data for this example are extracted from the article "Development of a General Equation to Determine the Transfer Factor Feed-to-Meat for Radiocesium on the Basis of the Body Mass of Domestic Animals" by Nalezinski, et al (Health Phys. 1996). A scatter plot of transfer factors (Y) versus body mass (X) for 26 animals (monogasters – Calves, Pigs, Laying Hens, Chicken) is shown in Fig. 6.11. It appears that there is a monotonically decreasing relationship between transfer factors and body mass. The Spearman's Rho for these data is calculated to be –0.836. A test of the hypotheses

Null Hypothesis: X and Y are statistically independent of each other

Alternative Hypothesis: There is a negative monotonic relationship between X and Y

is calculated as follows. The entry in Table A10 of Conover (1999) with n = 26 and $p = 1-\alpha$ = 0.95 is 0.3299. Hence the critical value Rho1* = –0.3299. Since the value of Rho is smaller than this critical value, the null hypothesis is rejected and we conclude that the data provide evidence supporting the alternative hypothesis. The P-value corresponding to Rho = –0.836 (n=26) is calculated as the area to the left of

$Z = -(0.836)(\sqrt{25}) = -4.18$ under the standard normal curve. This area is practically zero, so the P-value is zero (to 4 decimal places). This establishes a strong negative association between transfer factor and body mass.

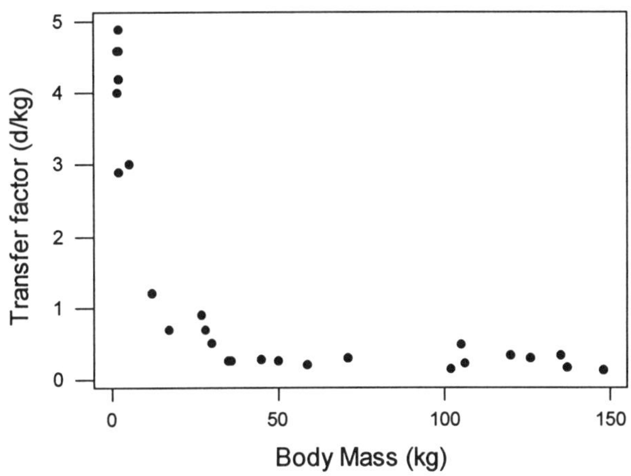

Figure 6.11. Scatter plot of transfer factor (d/kg) versus body mass (kg) in the radiocesium study.

We now take a closer look at the situation. Fig. 6.12 shows a plot of transfer factor versus body mass for animals weighing 50 kg or more. Based on this plot there appears to be no association between Y and X when X is restricted to the range 50 kg or greater. In fact, the Spearman's rho for this restricted data set is Rho = –0.123 and the P-value for a test of association is between Y and X is 0.349, a nonsignificant value.

For sake of completeness let us also look at the relationship between Y and X for the body mass range 0 kg to 50 kg. A scatter plot of the data in this range is shown in Fig. 6.13. There is definitely an association between Y and X here. The value of Spearman's Rho is –0.905 with a P-value of 0.0002.

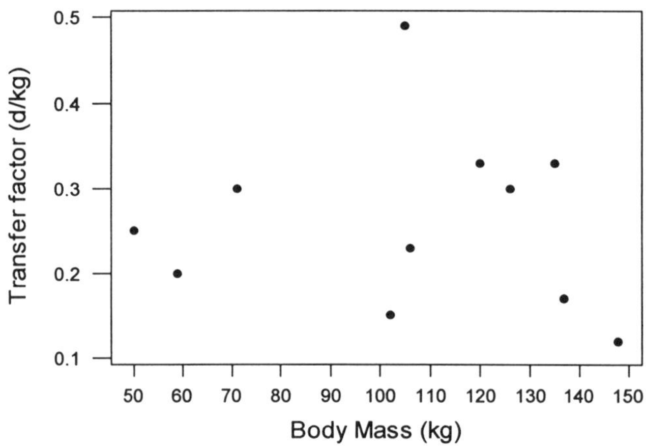

Figure 6.12. Scatter plot of transfer factor (d/kg) versus body mass (kg) in the radiocesium study for animals weighing 50 kg or more.

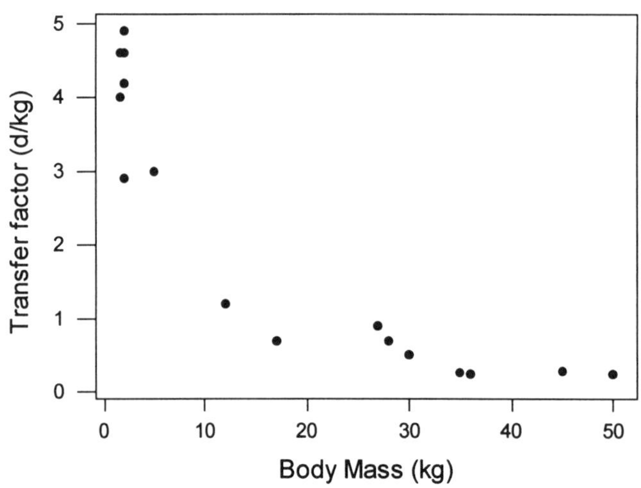

Figure 6.13. Scatter plot of transfer factor (d/kg) versus body mass (kg) in the radiocesium study for animals weighing 50 kg or less.

When there is a strong association between X and Y, one often attempts to describe this association using a suitable equation relating X and Y. Linear and Nonlinear regression analyses may be used to further study these data. Sometimes a suitable transformation simplifies the modeling process. For the data in the body mass range 0 kg – 50 kg we have plotted the logarithms (base 10) of the transfer factors versus body mass. See Fig. 6.14. The relationship between Y and X in this plot appears to be linear. A least squares straight line fit will most likely provide a very good approximation for the relationship between $\log_{10}(Y)$ and X in the range from 0 kg – 50 kg.

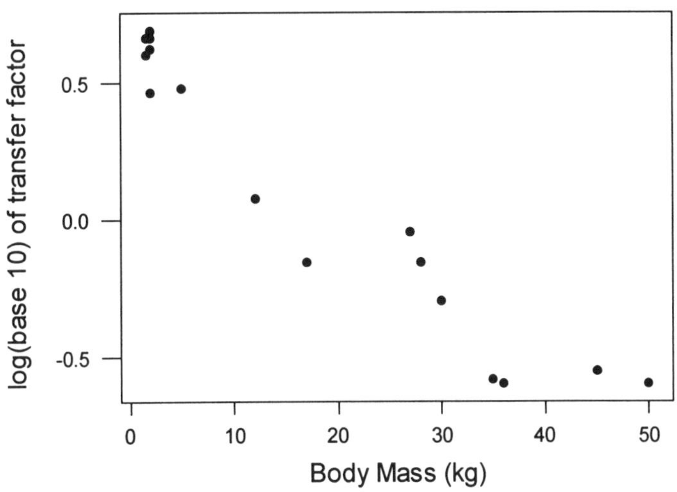

Figure 6.14. Scatter plot of the logarithms (base 10) of transfer factor (d/kg) versus body mass (kg) in the radiocesium study for animals weighing 50 kg or less.

6.11 SEN'S SLOPE ESTIMATOR FOR STRAIGHT LINE REGRESSION

When the relationship between Y and X appears to be adequately modeled using a straight line, it is of interest to estimate the slope of the straight line since it carries the information about the average rate at which Y is changing as a function of X. The method of least squares (see Chapter 5) is considered to be the preferred method for estimating the slope of the straight line if the underlying data are normally distributed. When normality assumption does not hold, various nonparametric counterparts to the least squares method may be considered.

6.11.1 Sen's Slope Estimator

A popular nonparametric estimator of slope is the one proposed by Sen (1968), which is an extension of a procedure originally proposed by Theil (1950). According to Sen's method, one considers all pairs of points $(X_i, Y_i), (X_j, Y_j)$ $(i < j)$ and computes a slope estimate

$$Slope = \frac{Y_i - Y_j}{X_i - X_j} \tag{6.17}$$

from each pair. We denote the total number of slope estimates by N^*. Assuming that there are n data pairs and that there are no ties in the X-values, the value of N^* is $n(n-1)/2$. If there are ties in the X values, then N^* is smaller than $n(n-1)/2$. The median of these pairwise slopes is Sen's estimate of the true slope β of the straight line regression of Y on X.

Here we illustrate Sen's method. We use the transfer factor versus body mass data from the previous section where body mass is restricted to the weight range 0 kg – 50 kg, and work with the logarithms (base 10) of the transfer factors. The number of possible pairwise slopes is 116 rather than the number $n(n-1)/2 = 120$ because of ties (there are 2 values of X equal to 1.6 each and 3 values of X equal to 2 each). The list of all 116 pairwise slopes can be calculated using MINITAB. The result is shown in Exhibit 6.10 where the values have been sorted in increasing order. The median of these 116 pairwise slopes is equal to –0.0289. The reader may wish to verify that this estimate is very close to the estimated slope using the method of least squares (least squares estimated slope = -0.0286).

Exhibit 6.10. All pairwise slopes (116 slopes) for the transfer factors example (sorted from the smallest to the largest).

-0.66787	-0.46554	-0.10914	-0.09877	-0.08509	-0.07306	-0.07102
-0.06741	-0.06188	-0.06181	-0.06145	-0.06110	-0.05836	-0.05685
-0.05680	-0.05634	-0.05611	-0.05589	-0.05460	-0.05451	-0.05441
-0.05309	-0.05267	-0.05188	-0.05028	-0.05017	-0.04915	-0.04871
-0.04682	-0.04088	-0.03864	-0.03801	-0.03794	-0.03781	-0.03736
-0.03720	-0.03677	-0.03675	-0.03661	-0.03604	-0.03554	-0.03540
-0.03540	-0.03500	-0.03481	-0.03442	-0.03394	-0.03301	-0.03250
-0.03180	-0.03164	-0.03145	-0.03122	-0.03113	-0.03097	-0.02993
-0.02944	-0.02891	-0.02888	-0.02867	-0.02839	-0.02834	-0.02827
-0.02817	-0.02801	-0.02789	-0.02748	-0.02735	-0.02717	-0.02692
-0.02676	-0.02661	-0.02635	-0.02613	-0.02575	-0.02553	-0.02550
-0.02488	-0.02419	-0.02398	-0.02390	-0.02377	-0.02365	-0.02356
-0.02353	-0.02341	-0.02213	-0.02112	-0.02033	-0.02025	-0.01915
-0.01793	-0.01703	-0.01679	-0.01505	-0.01463	-0.01421	-0.01355
-0.01124	-0.00984	-0.00833	-0.00114	0.00000	0.00000	0.00000
0.00322	0.00475	0.00547	0.01091	0.05297	0.06860	0.15174
0.22034	1.60851	2.00360	2.27798			

6.11.2 A Nonparametric Confidence Interval for Slope

There are also statistical significance tests and confidence intervals based on all pairwise slopes. We now describe the computation of a confidence interval for the true slope β (Gilbert, 1987).

1. Select a confidence level $1-\alpha$ (usually $1-\alpha$ = 0.95 or 0.99, corresponding to α=0.05 or 0.01, respectively).
2. Suppose the number of distinct Y values is R and d_j is the number of tied Y values in the jth group of tied Y values, $j=1,2,\cdots,R$. Also let S denote the number of distinct X values and u_k denote the number of tied X values in the kth group of tied X values, $k=1,2,\cdots,S$. Compute $C=Z_{1-\alpha/2}V$ where n is the sample size, $Z_{1-\alpha/2}$ is the $1-\alpha/2$ quantile of the standard normal distribution, and

$$V = \frac{1}{18}[n(n-1)(2n+5) - \sum_{j=1}^{R} d_j(d_j-1)(2d_j+5) - \sum_{k=1}^{S} u_k(u_k-1)(2u_k+5)]$$
$$+ \frac{[\sum_{j=1}^{R} d_j(d_j-1)(d_j-2)][\sum_{k=1}^{S} u_j(u_j-1)(u_j-2)]}{9n(n-1)(n-2)} + \frac{[\sum_{j=1}^{R} d_j(d_j-1)][\sum_{k=1}^{S} u_k(u_k-1)]}{2n(n-1)} \quad (6.18)$$

When there are no ties in the X values, the expression for V simplifies and we get,

$$V = \frac{1}{18}[n(n-1)(2n+5) - \sum_{j=1}^{R} d_j(d_j-1)(2d_j+5)] \quad (6.19)$$

When there are no ties in either the X values or the Y values, the expression for V becomes

$$V = \frac{1}{18}[n(n-1)(2n+5)] \quad (6.20)$$

3. Let $M_1 = (N^* - C)/2$ and $M_2 = (N^* + C)/2$.
4. Order the N^* pairwise slopes from the smallest to the largest.
5. Let b_L denote the M_1^{th} of the N^* ordered pairwise slopes counting up from the smallest to the largest. This is the lower limit of a two-sided $1-\alpha$ confidence interval for the true slope. If M_1 is not an integer, then round it down to an integer (or use interpolation).
6. Let b_U denote the $(M_2+1)^{th}$ of the N^* ordered pairwise slopes counting up from the smallest to the largest. This is the upper limit of a two-sided $1-\alpha$ confidence interval for the true slope. If (M_2+1) is not an integer, then round it up to an integer (or use interpolation).
7. A two-sided (approximate) $1-\alpha$ confidence interval for the true slope is $b_L \le \beta \le b_U$.

6.11.3 A Nonparametric Test for Slope

The confidence interval procedure given above can be used to construct hypothesis tests about β. As usual, we consider three scenarios. In the following, β_0 denotes a reference value to which the true slope will be compared. In applications β_0 is often zero.

*6.11.3.1 **Left-sided Alternative.*** Suppose we wish to test the following hypotheses at level α .
Null Hypothesis: β is equal to β_0
Alternative Hypothesis: β is less than β_0

The test is carried out as follows. Compute the upper limit b_U of a $1-2\alpha$ two-sided confidence interval for β. The null hypothesis is rejected in favor of the alternative hypothesis if b_U is less than β_0.

*6.11.3.2 **Right-sided Alternative.*** Suppose we wish to test the following hypotheses at level α .
Null Hypothesis: β is equal to β_0
Alternative Hypothesis: β is greater than β_0

The test is carried out as follows. Compute the lower limit b_L of a $1-2\alpha$ two-sided confidence interval for β. The null hypothesis is rejected in favor of the alternative hypothesis if b_L is greater than β_0.

*6.11.3.3 **Two-sided Alternative.*** Suppose we wish to test the following hypotheses at level α .
Null Hypothesis: β is equal to β_0
Alternative Hypothesis: β is not equal to β_0

The test is carried out as follows. Compute the lower limit b_L and the upper limit b_U of a $1-\alpha$ two-sided confidence interval for β. The null hypothesis is rejected in favor of the alternative hypothesis if β_0 is not contained in the interval from b_L to b_U.

We illustrate the computations using the transfer factors data for animals weighing below 50 kg.

6.11.4 Illustrative Example for a Nonparametric Test of Slope

Consider the transfer factors data for animals weighing less than 50 kg. We will use the logarithmic transformation (base 10) of the transfer factor in our analysis. Suppose we wish to test, using $\alpha = 0.05$, the pair of hypotheses

Null Hypothesis: β is equal to 0

Alternative Hypothesis: β is less than 0

We need to compute the upper limit b_U of a confidence interval for β with confidence level $1-2\alpha = 0.90$. Observe that there are tied values in both X and Y. For Y, there are 2 tied values with each equal to 0.25, two tied values with each equal to 0.70, and two tied values with each equal to 4.60. Therefore, the number of distinct Y values is $R = 13$ and $d_1 = 2,\ d_2 = 2,\ d_3 = 2$, and d_4 through d_{13} are all equal to 1. For X, there are two tied value with each equal to 1.6 and 3 tied values with each equal to 2.0. Thus $S = 13$, $u_1 = 2, u_2 = 3$ and all the remaining u_k, $k = 3, \cdots 13$, are equal to 1. Using these values we get $V = 485.767$. Furthermore, the normal table value that is needed is $Z_{1-(2\alpha)/2} = Z_{0.95} = 1.645$. So $C = (1.645)(485.767)^{1/2} = = 36.2560$. Also, $N^*=116$, so $M_1 = (N^*-C)/2 = 39.8720 = 39$ (rounded down) and $M_2 = (N^*+C)/2 = 76.1280 = 77$ (rounded up). Sorting the 116 pairwise slopes and selecting the 77th slope (counting from the smallest) we get $b_U = -0.0255$. Hence the null hypothesis is rejected and we conclude that the true slope is less than zero.

Sen's slope estimate should be used only when a straight line model appears to be a reasonable model for describing the relationship between Y and X. For instance, it should not be used to estimate a slope for the data plotted in Fig. 6.9 or in Fig. 6.10, because the relationship between Y and X is not linear (although the relationship in each case is perfectly monotonic). Even when normality assumptions hold, one might wish to consider using the Sen's slope estimator because it is more robust to outliers than the usual least squares estimator.

6.12 FINAL REMARKS

In this chapter we presented an overview of some of the most commonly used nonparametric methods and illustrated these methods using examples from the Health Physics journal. Formulas were presented for most computations. Concurrently we also discussed how the computations may be performed using MINITAB. For more advanced topics in nonparametric methods, the reader may refer to Conover (1999) or Hollander and Wolfe (1999).

6.13 REFERENCES

Bose, S. R.; Williamson, T. G.; Mulder, R. U.; Rab Molla, M. A. Impact of a 2 MW_{th} research reactor on radioactivity in sediments. Health Physics: 65:200-208; 1993.

Conover, W. J. Practical nonparametric statistics. 2nd Edition. New York: Wiley; 1999.

Gilbert, R. O. Statistical methods for environmental pollution monitoring. New York: Van Nostrand Reinhold, 1987.

Hollander, M.; Wolfe, D. A. Nonparametric Statistical Methods. 2nd Edition. New York: Wiley; 1999.

Ludwig, T.; Schwass, D.; Seitz, G.; Siekmann, H. Intakes of thorium while using thoriated tungsten electrodes for TIG welding. Health Physics. 77:462-469; 1999.

Minitab Inc. MINITAB user's guide 1 & 2. Release 12, 1997.

Nalezinski, S.; Ruhm, W.; Wirth, E. Development of a general equation to determine the transfer factor feed-to-meat for radiocesium on the basis of the body mass of domestic animals. Health Physics. 70:717-721; 1996.

Sen, P. K. Estimates of the regression coefficient based on Kendall's tau. Journal of the American Statistical Association. 63:1379-1389; 1968.

Thiel, H. A rank-invariant method of linear and polynomial regression analysis. Indagationes Mathematicac, 12:85-91, 1950.

6.14 AUTHOR PROFILE

Hari Iyer is a Professor of statistics at Colorado State University. He received a Ph.D in mathematics from the University of Notre Dame and a Ph.D. in statistics from Colorado State University. His research interests include linear models, experimental design, permutation methods, environmental statistics and pharmaceutical statistics.

Table A. Critical values for the sign test. $m =$ effective sample size. Values given are the critical values T^* for one-sided alternatives, corresponding to $\alpha = 0.05$, $\alpha = 0.025$, $\alpha = 0.01$, and $\alpha = 0.005$. For an α level test with a two-sided alternative, use the $\alpha/2$ column to obtain the critical value T^*.

	α					α					α					α			
m	0.05	0.025	0.01	0.005	m	0.05	0.025	0.01	0.005	m	0.05	0.025	0.01	0.005	m	0.05	0.025	0.01	0.005
3	0	0	0	0	**40**	15	14	13	12	**77**	31	30	28	27	**114**	48	47	45	43
4	0	0	0	0	**41**	15	14	13	12	**78**	32	30	29	28	**115**	49	47	45	44
5	1	0	0	0	**42**	16	15	14	13	**79**	32	31	29	28	**116**	49	47	46	44
6	1	1	0	0	**43**	16	15	14	13	**80**	33	31	30	29	**117**	50	48	46	45
7	1	1	1	0	**44**	17	16	14	14	**81**	33	32	30	29	**118**	50	48	46	45
8	2	1	1	1	**45**	17	16	15	14	**82**	34	32	31	29	**119**	51	49	47	46
9	2	2	1	1	**46**	17	16	15	14	**83**	34	33	31	30	**120**	51	49	47	46
10	2	2	1	1	**47**	18	17	16	15	**84**	34	33	31	30	**121**	51	50	48	46
11	3	2	2	1	**48**	18	17	16	15	**85**	35	33	32	31	**122**	52	50	48	47
12	3	3	2	2	**49**	19	18	16	16	**86**	35	34	32	31	**123**	52	51	49	47
13	4	3	2	2	**50**	19	18	17	16	**87**	36	34	33	32	**124**	53	51	49	48
14	4	3	3	2	**51**	20	19	17	16	**88**	36	35	33	32	**125**	53	52	50	48
15	4	4	3	3	**52**	20	19	18	17	**89**	37	35	34	32	**126**	54	52	50	49
16	5	4	3	3	**53**	21	19	18	17	**90**	37	36	34	33	**127**	54	52	50	49
17	5	5	4	3	**54**	21	20	19	18	**91**	38	36	34	33	**128**	55	53	51	49
18	6	5	4	4	**55**	21	20	19	18	**92**	38	37	35	34	**129**	55	53	51	50
19	6	5	5	4	**56**	22	21	19	18	**93**	39	37	35	34	**130**	56	54	52	50
20	6	6	5	4	**57**	22	21	20	19	**94**	39	38	36	35	**131**	56	54	52	51
21	7	6	5	5	**58**	23	22	20	19	**95**	39	38	36	35	**132**	57	55	53	51
22	7	6	6	5	**59**	23	22	21	20	**96**	40	38	37	35	**133**	57	55	53	52
23	8	7	6	5	**60**	24	22	21	20	**97**	40	39	37	36	**134**	57	56	54	52
24	8	7	6	6	**61**	24	23	21	21	**98**	41	39	38	36	**135**	58	56	54	53
25	8	8	7	6	**62**	25	23	22	21	**99**	41	40	38	37	**136**	58	57	54	53
26	9	8	7	7	**63**	25	24	22	21	**100**	42	40	38	37	**137**	59	57	55	53
27	9	8	8	7	**64**	25	24	23	22	**101**	42	41	39	38	**138**	59	58	55	54
28	10	9	8	7	**65**	26	25	23	22	**102**	43	41	39	38	**139**	60	58	56	54
29	10	9	8	8	**66**	26	25	24	23	**103**	43	42	40	38	**140**	60	58	56	55
30	11	10	9	8	**67**	27	26	24	23	**104**	44	42	40	39	**141**	61	59	57	55
31	11	10	9	8	**68**	27	26	24	23	**105**	44	42	41	39	**142**	61	59	57	56
32	11	10	9	9	**69**	28	26	25	24	**106**	45	43	41	40	**143**	62	60	58	56
33	12	11	10	9	**70**	28	27	25	24	**107**	45	43	42	40	**144**	62	60	58	57
34	12	11	10	10	**71**	29	27	26	25	**108**	45	44	42	41	**145**	63	61	59	57
35	13	12	11	10	**72**	29	28	26	25	**109**	46	44	42	41	**146**	63	61	59	57
36	13	12	11	10	**73**	29	28	27	26	**110**	46	45	43	42	**147**	64	62	59	58
37	14	13	11	11	**74**	30	29	27	26	**111**	47	45	43	42	**148**	64	62	60	58
38	14	13	12	11	**75**	30	29	27	26	**112**	47	46	44	42	**149**	64	63	60	59
39	14	13	12	12	**76**	31	29	28	27	**113**	48	46	44	43	**150**	65	63	61	59

Chapter 7.0

DEVELOPING FIELD SAMPLING PLANS USING THE DATA QUALITY OBJECTIVES PROCESS

Richard O. Gilbert and John E. Wilson

7.1 INTRODUCTION

This chapter provides guidance on how to use the Data Quality Objectives (DQO) process to develop sampling designs for final status radionuclide surveys of soils and buildings to demonstrate compliance with dose or risk-based standards. The process is discussed and illustrated in the context of the Multi-Agency Radiation Survey and Site Investigation Manual (MARSSIM) (USNRC 1997) and with the aid of VSP (Visual Sample Plan) design software. Hence, this chapter provides not only a knowledge of the DQO process, but also an understanding of the measurement and sampling designs in USNRC (1997) as well as why and how they were selected.

The learning objectives of this chapter are:

1. Understand the logic and usefulness of the DQO process.
2. Understand the major design options and how to select one.
3. Understand the designs used in MARSSIM and why they were selected.
4. Obtain hands-on-experience using the Visual Sample Plan (VSP) software to help design, graphically display, and implement sampling strategies for a final status survey.

This chapter focuses on sampling designs for soil and buildings. Designs for sampling air, ground water, surface water, plants and animals are not emphasized due to space and resource limitations. The motivation for writing this chapter springs from a desire to avoid inefficient sampling designs that arise from the following types of problems:

- Not clearly understanding or defining the problem that motivates the measurement/sampling effort and the decisions that need to be made using those data. Why are data needed and how will they be used?
- Taking measurements and collecting samples without adequate thought and planning. Are these the data what we really need to make the necessary

decisions? Are they the correct type and are they of sufficient quality and quantity?

- Not using statistical formulas to supplement our knowledge of nuclear facility operations and past data gathering efforts to determine the optimal type, number and location of measurements and samples.
- Not using statistical methods to take into account our tolerance for decision and estimation errors when determining the type, quality and quantity of measurements and samples.

The DQO process is introduced in Section 7.2 and illustrated with a MARSSIM case study in Section 7.3. The major sampling designs are discussed and compared in Section 7.4. Section 7.5 focuses on the number of samples required for a sampling design. Section 7.6 describes the Visual Sample Plan (VSP) software package that provides tools for selecting and visually displaying a sampling design and for computing the number of samples required using outputs from the DQO process. References are listed in Section 7.7.

7.2 THE DATA QUALITY OBJECTIVES PROCESS

The Data Quality Objectives (DQO) process is a 7-step, structured planning process developed by the U.S. Environmental Protection Agency (USEPA 1993, 1994, 1999) for obtaining data that will be sufficient for their intended use. The DQO process is designed to answer the following questions:

- What do you need?
- Why do you need it?
- How will you use it?
- What is your tolerance for decision errors?

No radionuclide measurements or samples should be collected without first answering these questions. Otherwise, it is much more likely that data of the wrong type, quality or quantity obtained at inappropriate times and locations will be obtained.

The DQO process is a structured approach for asking the right questions and using the answers to develop an efficient sampling plan. The DQO process is a tool to balance cost versus uncertainty. As we can allow more uncertainty in decision making, we decrease our costs because fewer samples and measurements are needed. But as our tolerance for uncertainty and making decision errors decreases, then either our costs will increase in order to obtain more data, or we will need to use a more efficiently sampling and measurement strategy.

The DQO process is flexible and should be used iteratively through the various phases of a multi-phase project (e.g., historical site assessment, scoping survey, characterization survey, remediation, remedial action support survey, and final status survey). The DQO process should be applied to the ultimate decision (e.g., "Can the site be released for unrestricted use?") as well as to each phase in turn to make sure the appropriate information and data are obtained in all phases. This philosophy is used in USNRC (1997, Chapter 5). The DQO process can also be iterative within each phase of a project. For example, during the final status survey, we may try out several different combinations of cost and tolerable levels of uncertainty before we strike an acceptable balance.

The DQO process is the planning step in the overall project life cycle, which consists of planning, implementation (collecting data) and assessment (assuring the correct type, quality and quantity of data were collected). The Data Quality Assessment (DQA) process (USEPA 1998b) is used to conduct the assessment. Once the data are verified as having the desired quality, which includes verifying that the assumptions underlying statistical tests and other analyses are valid, the data can be used to make decisions.

The DQO process has seven steps. Each step and the activities associated with each step are summarized in Table 7.1 and illustrated in the case study described in Section 7.3.

The first six steps are needed to achieve Step 7 (optimize the sampling design). Steps 1 through 6 are the "homework" needed to assure that the data obtained will be representative and of the type, quality, and quantity needed to achieve the goals of the sampling effort. However, several iterations between Steps 6 and 7 (or Steps 1-5) may be needed before a final design is selected. This is illustrated in the case study in Section 7.3. The outputs of each step are data quality objectives (DQOs), which are qualitative and quantitative criteria that clarify study objectives, define appropriate types of data to collect and specify the tolerable levels of potential decision errors.

Table 7.1. Steps and activities in the data quality objectives (DQO) process

Step 1: State the Problem

Purpose: Define the problem so the focus of the study will be unambiguous

Activities:

- Identify the members of the planning team and stakeholders
- Identify the primary decision maker or decision-making method
- Develop a concise description of problem
- Specify the available resources and relevant deadlines for the study

Step 2: Identify the Decision

Purpose: Define the specific decisions that need to be made or the questions that need to be answered

Activities:

- Identify the principal study question
- Define the alternative actions that could result from a resolution of the principal study question
- Combine the principal study question and the alternative actions into a decision statement
- Organize multiple decisions

Step 3: Identify Inputs to the Decision

Purpose: Identify the data or other information required to resolve the decision statement

Activities:

- Identify the information required to resolve the decision statement
- Identify and list the sources of required information of each item identified
- Identify the information needed to establish the action level or the derived concentration guideline level (DCGL)
- Confirm that appropriate field and laboratory measurement methods exist to provide the necessary data

Step 4: Define the Study Boundaries

Purpose: Define the spatial and temporal boundaries that are covered by the decision statement

Activities:

- Specify the characteristics that define the population of interest
- Define the geographical area within which all decisions must apply
- If appropriate, divide the site into areas or survey units that have relatively homogeneous characteristics
- Determine the time frame to which the decision applies
- Determine when to collect data
- Define the scale of decision making
- Identify practical constraints on data collection

Step 5: Develop a Decision Rule

Purpose: Define an "if...then" decision rule that will specify how data will be used to make a decision

Activities:

- Specify the statistical parameter of interest (typically the mean, median or an upper percentile) for making decisions
- Specify the action level for the study
- Combine the outputs of DQO steps 1-4 into an "if...then" decision rule that defines the conditions that would cause the decision-maker to choose among alternative actions.

Step 6: Specify Limits on Decision Errors

Purpose: Specify the decision maker's tolerable limits on the probability of making decision errors

Activities:

- Determine using professional judgement the possible range (upper and lower bounds) of the statistical parameter of interest
- Specify tolerable decision error (Type I and Type II) rates and choose the null hypothesis
- Specify the range (gray region) of possible statistical parameter values where the consequences of decision errors are relatively minor
- Assign probability limits on decision errors to values of the parameter that are above and below the gray region

Step 7: Optimize the Design for Obtaining Data
Purpose: Identify a resource-effective data collection design for generating data that are expected to satisfy the DQOs
Activities:

- Review DQO outputs of Steps 1-6 and existing environmental data to ensure they are internally consistent
- Develop general data collection design alternatives
- For each data collection design strategy determine the appropriate number of samples that will satisfy the DQO outputs
- Select the most resource-effective design that satisfies all the DQOs
- Document the operational details and theoretical assumptions of the selected design in the Quality Assurance Project Plan (QAPP), the field sampling plan, the sampling and analysis plan, or the decommissioning plan

7.3 MARSSIM CASE STUDY

In this section a final status survey at the Cushing Refinery site in Cushing, Oklahoma is used to illustrate the DQO process. This case study is from Abelquist (1997) and is also discussed in Annex I of IAEA (1999). Please refer to Table 7.1 as you proceed through the 7 steps of this case study.

7.3.1 DQO Step 1: State the Problem.

Kerr-McGee Corporation (KMC) operated an oil refinery at the Cushing site (~178 hectares) from 1956 to 1972. KMC used part of the Cushing site to process natural thorium and natural, depleted, and enriched uranium. In April 1966 KMC reported to the U.S. Atomic Energy Commission that all special nuclear material had been transferred from the Cushing site to KMC's new Cimarron facility in Crescent, Oklahoma and that all Cushing buildings in which licensed activities had been performed were cleaned and decontaminated.

This case study focuses on two portions of the site: a 2,300 square meter portion called RMA-4 and a reference (background) area for RMA-4. In 1996 the Environmental Survey and Site Assessment Program of the Oak Ridge Institute for Science and Education conducted site surveys at these two (as well as other) sites to demonstrate the feasibility of implementing the MARSSIM methodology at this site, which is contaminated with thorium. The results of this evaluation were published in Abelquist (1997).

The "problem" being faced is whether ^{232}Th concentrations in surface soil at the RMA–4 site are sufficiently low that the site can be released for unrestricted public use.

Members of the DQO planning team could include but not be limited to:

- the final status survey project leader
- someone with experience facilitating and implementing the DQO process
- someone familiar with historical operations at the site
- an expert in radiological survey instruments and analytical methods
- someone who is familiar with past RMA-4 and reference (background) survey data, including field measurement and sampling methods, quality control, radiation detectors and laboratory analytical methods
- a statistician or environmental professional who has knowledge and experience with the statistical design and analysis of environmental radionuclide surveys and data
- a risk assessor
- someone who is familiar with Nuclear Regulatory Commission (NRC) procedures and guidelines
- a regulator from the State of Oklahoma
- the decision maker

Available resources and deadlines for completing the RMA-4 final status survey were determined. A decision maker(s) is selected.

7.3.2 Step 2: Identify the Decision

The principal study question is whether the residual radioactivity levels of ^{232}Th in soil at the RMA-4 site meet the release criterion above background. Resolution of the principal study question could result in the following alternative actions:

- The RMA-4 site can be released for unrestricted use by the public.
- The RMA-4 site cannot be released for unrestricted use by the public and additional information is needed to determine how and when a restricted or unrestricted release can be achieved.

The decision statement is:

- Determine whether residual radioactivity levels of ^{232}Th in surface soil at the RMA-4 site are above the release criterion. If so, the site can be released for unrestricted use by the public; otherwise, additional information is needed to assess what actions are needed to achieve unrestricted release or a limited use release.

7.3.3 Step 3: Identify Inputs to the Decision

Information required to resolve the decision statement include but are not limited to:

- Historical site operations, reports, data, maps or photographs of the site
- The radionuclides of concern. For this case study ^{232}Th is the only radionuclide of concern.
- Information needed to determine action levels. This information includes regulatory guidance, dose-transport models, type of in-situ detectors and analytical measurement methods that will be used, the spacing between soil sampling locations in the field, and the size of any areas of elevated concentrations that are found.
- Determination of whether the RMA-4 site is a "Class 1," "Class 2," or "Class 3" site, as defined in USNRC (1997, page 2-5). RMA-4 is a Class 1 site, i.e., it is a site that has the highest potential for radioactive contamination above the derived concentration guideline level for the average concentration, denoted by $DCGL_W$ (discussed in Step 5 below), including the potential for small areas of elevated activity. Knowledge of the $DCGL_W$ is needed to determine the class designation (1, 2, or 3)
- An appropriate reference (background) area
- Detector survey instrumentation and soil sampling collection and measurement methods, including specification of required detection limits
- Surface soil scans for gamma radiation in RMA-4 and the reference area using NaI scintillation detectors
- ^{232}Th measurements of surface soil samples (0 to 15 cm) collected from RMA-4 and the reference area

- Inputs needed to determine the number of soil samples and their locations at RMA-4 and the reference area (see Step 6 and Step 7)

7.3.4 Step 4: Define the Boundaries

There are two populations of interest within the spatial and temporal boundaries of RMA-4 and the reference area:

- the set of all possibie surface (0-15 cm) soil samples that could be collected and
- the set of all possible surface scan NaI detector readings that could be obtained.

The geographical area (spatial boundary) of the RMA-4 and reference sites are shown in Figures 2 and 4, respectively, of Abelquist (1997, pages 26 and 28). The temporal boundary of RMA-4 and the reference area is the time period during which detector scans and soil samples will be collected and measured. The time frame to which the decision applies extends from the time of scanning and soil sampling until some point in the future when the site becomes contaminated again.

The scale of decision making is the survey unit, which is defined in USNRC (1997, page 2-4) to be a physical area consisting of structure or land areas of specified size and shape for which a separate decision will be made as to whether or not that area exceeds the release criterion. The survey unit in this example is the entire RMA-4 area. That is, a single decision will be made for the entire RMA-4 site (approximately 2300 m^2) rather than, for example, making a separate decision for each of several smaller subareas within RMA-4. Practical constrains on data collection may include availability of field crews and radiation detectors, budget problems, inclement weather, difficulty in obtaining access to the site, completion of the sampling and analysis plan, etc.

7.3.5 Step 5: Develop a Decision Rule

This step requires 3 activities.

7.3.5.1 ***Specify the Statistical Parameter of Interest for Making Decisions.*** The decision of whether RMA-4 can be released for unrestricted use by the public will depend on:

- the true *average* concentration of ^{232}Th

- the size of areas of elevated concentrations of ^{232}Th that are missed by soil samples collected on a systematic grid but that will be detected by NaI detector scans over the survey unit.

7.3.5.2 ***Specify Action Levels.*** We need to specify an action level for both of these parameters (the average concentration and the area size). The philosophy used in USNRC (1997, 1998) is to use the DQO planning process primarily for determining the sampling design for testing whether the true average concentration exceeds the action level $DCGL_W$. In this example, we illustrate that the DQO process can also be used to develop the plan for using NaI detector scans at RMA-4. The applicable action level [denoted as $DCGL_W$ in USNRC (1997, 1998)] for the true average concentration of ^{232}Th in surface soil samples for the entire RMA-4 site was stated in Abelquist (1997, page 5) to be 5.92×10^{-3} Bq g^{-1} (0.16 pCi g^{-1}) above background. That is:

$$DCGL_W = 5.92 \times 10^{-3} \text{ Bq g}^{-1} \tag{7.1}$$

This action level was developed on the basis of a specified release criterion expressed in terms of dose (mSv/y or mrem/y). The action level for ^{232}Th concentrations for *areas of elevated concentration* is denoted by $DCGL_{EMC}$, where EMC denotes elevated measurement comparison. The $DCGL_{EMC}$ is determined by going through the box-and-arrow diagram in Figure 5.3 of USNRC (1997, page 5-24). The first step in determining this action level is to determine the number of soil samples required to test whether the true average concentration of ^{232}Th exceeds $DCGL_W$ above background. This number of samples is shown in DQO Step 7 below (Section 7.3.7) to be 64. Also, it was decided in Step 7 that these 64 samples will be collected on a triangular grid over the RMA-4 site. (See Section 7.4.4 for an illustration of a triangular grid.) Hence, the spacing between the grid points of the triangular grid over the 2,300 m^2 RMA-4 area is 7.44 m (see Abelquist 1997, page 1), so the largest area of elevated ^{232}Th concentrations that could exist without being detected by measuring ^{232}Th in soil samples at the grid points is 36 m^2.

The next step in determining the action level for a 36 m^2 area of elevated concentrations is to determine the magnitude (area dose factor) that the concentration in that area can exceed the $DCGL_W$ value of 5.92×10^{-3} Bq g^{-1} while not exceeding the dose release criterion (in mSv/y or mrem/y). Using Table 7.2 below (from Table 3 in Abelquist 1997, page 11) we find that the area dose factor is 100. The area dose factors in Table 7.2 were determined using an exposure pathway model. As stated in USNRC

(1998, page 3-29), values for the area dose factor can be determined by comparing the dose conversion factor obtained from the results of a dose-pathway analysis under the scenario that a unit activity concentration of a given radionuclide is distributed uniformly across the survey unit to the dose conversion factor obtained when a unit concentration of that radionuclide is confined to the smaller elevated concentration area. See USNRC (1997, pages 5-36 through 5-39) and USNRC (1998, pages 3-29 through 3-31) for additional information on area dose factors.

Table 7.2 Outdoor area dose factors for ^{232}Th

Area (m^2)	3610	1200	361	120	36.1	5.54	2.32	1.47	1.00
Area Dose Factor	1	3	10	30	100	300	1000	3000	10000

Next, the area dose factor of 100 is used to compute the required minimum detectable concentration (MDC) of any NaI gamma detector that might be used at RMA-4:

$$\text{MDC required} = \text{DCGL}_W \times (\text{Area Dose Factor}) = 5.92\text{x}10^{-3}\ \text{Bq g}^{-1} \times 100 = 0.592\ \text{Bq g}^{-1} \quad (7.2)$$

Now, as discussed in Abelquist (1997, page 12), the MDC of the selected NaI scintillation detector that was to be used at the site was 0.14 Bq g^{-1}, which is less than the required 0.592 Bq g^{-1}. Hence, use of that detector over 100% of the RMA-4 area will detect any area of elevated concentration that would cause the release criterion for the area to be exceeded. Hence,

$$\text{DCGL}_{EMC} = 0.592\ \text{Bq g}^{-1} \quad (7.3)$$

Note that if the NaI detector used did not have a MDC less than the required MDC, then additional soil samples would be needed in order to reduce the spacing between soil samples collected on the triangular grid. This would decrease the size of the area of elevated concentration that would not be detected by the grid soil samples. Since the size of that area is smaller, the concentration within that area could be larger without exceeding the release criteria for the site, which would allow the use of a detector with a larger MDC. In that case, the $DCGL_{EMC}$ would become larger than 0.592 Bq g^{-1}. We note that USNRC (1998, pages 2-11 through 2-13) discusses the statistical hypothesis testing concepts that underlie the calculation of the MDC. A discussion of the derivation of MDC values is beyond the scope of this chapter.

In conclusion, *before* soil sampling and NaI scans were conducted at the RMA-4 site, the two action levels are $DCGL_W$ = 5.92×10^{-3} Bq g^{-1} and $DCGL_{EMC}$ = 0.592 Bq g^{-1}. However, the *actual* size of any elevated areas less than 36 m^2 in size will not be known until the NaI scans are conducted. Some of the NaI scans conducted at the site exceeded the $DCGL_{EMC}$ = 0.592 Bq g^{-1}. The size of the elevated area was about 300 m^2, which was confirmed by collecting soil samples in that area. Now we must determine a *new* $DCGL_{EMC}$, one that corresponds to the *actual* area of elevated concentration. The new $DCGL_{EMC}$ was determined by multiplying the $DCGL_W$ (5.92×10^{-3} Bq g^{-1}) by 5.54, which is the area dose factor that corresponded to the 300 m^2 area. That is,

$$DCGL_{EMC} \text{ for } 300 \text{ m}^2 \text{ area} = DCGL_W \times (\text{Area Dose Factor}) = 5.92 \times 10^{-3} \text{ Bq g}^{-1} \times 5.54 = 3.38 \times 10^{-2} \text{ Bq g}^{-1} \quad (7.4)$$

The decision about whether to remediate the 300 m^2 area would depend on whether any individual measurement or the average of a suitable number of measurements (determined via the DQO process) in that area exceeds 3.38×10^{-2} Bq g^{-1}.

Suppose there are two (or more) areas of elevated concentrations. For example, Abelquist (1997, page 17) indicates that a particularly significant location of elevated concentration of size 20 m^2 was also found using the NaI scans. The $DCGL_{EMC}$ must also be computed for that size area, as follows:

$$\begin{aligned} DCGL_{EMC} \text{ for } 20 \text{ m}^2 \text{ Area} &= DCGL_W \times (\text{Area Dose Factor}) \\ &= 5.92 \times 10^{-3} \text{ Bq g}^{-1} \times 208 = 1.23 \text{ Bq g}^{-1} \end{aligned} \quad (7.5)$$

where the area dose factor (208) was determined by Abelquist (1997) by interpolating in his Table 3 (Table 7.2 above). The decision about whether to remediate the 20 m^2 area would depend on whether any individual measurement or the average of a suitable number of measurements (determined via the DQO process) in that area exceeds 1.23 Bq g^{-1}.

7.3.5.3 Develop an "if…then" decision rule. Combining the alternative actions from Step 2 and the scale of decision making from Step 4 with the statistical parameters and the action levels derived about in this step, we obtain the following decision rule.

- **DECISION RULE:** If the true mean concentration of ^{232}Th soil samples for the entire RMA-4 site exceeds $DCGL_W$ = 5.92x10^{-3} Bq g^{-1}, or if the true mean average of ^{232}Th concentrations within any area of elevated ^{232}Th soil concentrations exceeds the applicable $DCGL_{EMC}$ value (Bq g^{-1}), or if any individual ^{232}Th measurement of a soil sample or any single NaI detector measurement in that elevated area exceeds the applicable $DCGL_{EMC}$ value, then the RMA-4 site does not meet the dose release criterion, the site cannot be released for unrestricted use by the public, and additional investigations of the site are required; otherwise, the site can be released for unrestricted use by the public.

  This decision rule could be stated more succinctly as follows:

- If the true average concentration exceeds the $DCGL_W$ or if any combination of area and concentration exists that exceeds the dose release criteria, then the RMA-4 site does not meet the dose release criterion, the site cannot be released for unrestricted use by the public, and additional investigations of the site are required; otherwise, the site can be released for unrestricted use by the public.

We note that another decision rule may be formulated to determine when additional investigation is required for the purpose of quality control, e.g., to insure that the measurement process is not out of control or that the area was correctly classified as Class 1, 2, or 3. See USNRC (1997, page 5-44 through 5-46).

7.3.6 Step 6: Specify Limits on Decision Errors

This step requires four activities to determine in DQO Step 7 the number of soil samples that should be sampled and analyzed for ^{232}Th.

*7.3.6.1 **Determine the Plausible Range of the Statistical Parameters of Interest.*** For illustration purposes we shall assume the true average concentration of ^{232}Th above background in surface soil is not expected to be less than zero nor greater than 0.01 Bq g^{-1} (0.27 pCi g^{-1}).

7.3.6.2 Choose the Null Hypothesis and Identify (Define) Type I and Type II Decision Errors. The *null hypothesis* is the hypothesis the DQO planning team and decision maker decide to accept as being true unless the data strongly indicate otherwise. The null hypothesis recommended in USNRC (1997, page 2-26) is: "The residual radioactivity in the survey unit exceeds the release criterion," i.e., "guilty until proven innocent." This null hypothesis was used at the Cushing site, namely "The true mean ^{232}Th *exceeds* the $DCGL_W$ of 5.92×10^{-3} Bq g^{-1} above background." Hence, the burden on proof rests on showing that the true mean does *not* exceed the $DCGL_W$.

USNRC (1997, page 2-39) and USNRC (1998, pages 2-8 through 2-10) discuss when it may be suitable to use the opposite null hypothesis, i.e. that the residual radioactivity in the survey unit does *not* exceed the release criterion, i.e., "innocent until proven guilty." If this null hypothesis had been selected, the burden of proof would rest on showing that the true mean *exceeds* $DCGL_W$. It should be noted that the computations conducted on the collected data to test the null hypothesis depends on which null hypothesis is used. The different calculations are illustrated in USNRC (1998, page 2-20) for the Wilcoxon Rank Sum test, which is the test statistic that was selected for use at the Cushing site (see Section 7.3.7.3 below).

A *Type I decision error* is made when the null hypothesis is rejected when in fact the null hypothesis is true. For the null hypothesis above, a Type I decision error is the probability the true mean is accepted as being less than the $DCGL_W$ when in fact the true mean is greater than $DCGL_W$ (regulator's risk).

A *Type II decision error* is made when the null hypothesis is accepted when in fact it is not true. For this case study, a Type II decision error is made when true mean is accepted as being greater than the $DCGL_W$ when in fact the true mean is less than the $DCGL_W$.

Section 1.3.4 provides additional discussion about hypothesis testing.

7.3.6.3 Specify the range (gray region) of possible statistical parameter values where the consequences of decision errors are relatively minor. Since the null hypothesis is that the true mean $DCGL_W$ concentration in soil is greater than $DCGL_W$, the upper bound of the gray region is the mean concentration $DCGL_W$. The lower bound of the gray region (LBGR) is a mean concentration that is greater than or equal to zero but less than $DCGL_W$. It should be selected such that remediation when the true mean ^{232}Th concentration is less than or equal to the LBGR is not considered to be acceptable, achievable or reasonable from a cost and risk perspective (see USNRC 1998, pages 3-15 through 3-18 and USNRC 1997, pages 8-12, 8-13, 8-19, D-20). The gray region, i.e., the region between the LBGR and $DCGL_W$, contains potential values of the true mean that are less than $DCGL_W$. Hence, within the gray region the correct decision is that the site should indeed be released for unrestricted use by the public. Abelquist (1997, page 6) selected the LBGR to be zero.

- Hence the LBGR = 0 and the gray region extends from 0 (above background) to the $DCGL_W$ (5.92×10^{-3} Bq g^{-1} above background).

7.3.6.4 Assign probability limits on decision errors to potential values of the true mean that are above and below the gray region. Tolerable levels for the Type I error rate (α) and the Type II error rate (β) are typically determined by considering the consequences (health, ecosystem, political, social, practical) of making that decision errors. Abelquist (1997, page 5) selected $\alpha = 0.05$ (5 % error rate) and $\beta = 0.10$ (10 % error rate).

The meaning of these decision error rates is as follows:

If the true mean ^{232}Th concentration above background for the entire RMA-4 site should happen to be *less than or equal to the LBGR (0 Bq g^{-1})*, then we can tolerate no more than a 10% chance of deciding that the site should not be released for unrestricted use by the public. However, if the true mean ^{232}Th concentration above background for the entire RMA-4 site should happen to be *greater than or equal to the action level*, $DCGL_W$ (5.92×10^{-3} Bq g^{-1} above background), then we can tolerate at most a 5% chance of deciding that the site should be released for unrestricted use.

Factors to consider in setting probability values for α and β are discussed in USNRC (1998, pages 3-14 through 3-18) and USNRC (1997, pages D-21 through D-28).

7.3.7 Step 7: Optimize the Design for Obtaining Data

This step has 6 activities:

7.3.7.1 Review DQO Outputs of Steps 1-6 And Existing Environmental Data To Ensure They Are Internally Consistent. All outputs of Steps 1-6 should be internally consistent and should succinctly summarize information needed to develop the sampling and measurement design. It is a good idea to prepare a table that summarizes the outputs of the first 6 steps of the DQO process, as illustrated in USEPA (1999, page 42).

Review existing environmental data to see if there are data gaps. For example, are data available for estimating the standard deviation of ^{232}Th for the RMA-4 and background areas? This information will be needed below in this step of the DQO process.

7.3.7.2 Develop General Data Collection Design Alternatives. At this point, alternative designs for selecting soil samples over the RMA-4 site and the background site discussed above should be considered.

USNRC (1997, pages 2-25 through 2-40 and 5-21 through 5-55) provides a structure for determining the design (location and number of samples). The key features are:

- Put the burden of proof on demonstrating that the site can be released for unrestricted use, i.e., choose the H_o to be that the site *cannot* be released for unrestricted use.
- Divide the site into survey units for which separate release decisions are needed.
- Determine which survey units are potentially impacted and classify each unit as either Class 1 (for which exceedance of the release criterion is possible), Class 2 (for which exceedance is not expected), or class 3 (for which no or very little residual radioactivity is expected) (USNRC 1998, page 2-3; USNRC 1997 page 2-5).
- Place the most survey effort on Class 1 units, somewhat less on Class 2 units and still less on Class 3 units.
- Use nonparametric (distribution-free) statistical tests to make decisions to avoid having to know the probability distribution of the data.

- Design the measurement and sampling plan to permit conducting two types of statistical comparisons: (1) one appropriate for units that have relatively uniform distributions of contamination, and (2) one that checks for small areas of elevated activity.
- Determine the number of samples taking into consideration which type of statistical test will be used.
- Use a systematic square or triangular grid design if the site is a Class 1 or a Class 2 area. Use simple random sampling for a Class 3 area.

However, flexibility in applying this structured guidance is permitted (USNRC 1997, pages 2-33 through 2-40).

Alternative sampling designs that might be considered to obtain data to test whether median ^{232}Th surface soil concentrations in the RMA-4 area exceed the median for the background area at the Cushing site are described in Section 7.4 below. The designs discussed are judgement sampling, simple random sampling, square or triangular grid, stratified random sampling, double sampling, and ranked set sampling. The pros and cons of these designs are summarized in Table 7.2.

A triangular grid was used at both the RMA-4 and background area at the Cushing site because it provided for uniform coverage of those sites and there were no prohibitive cost or practical problems involved in implementing that design. However, if the objective had been to estimate the mean without needing to also look efficiently for areas of elevated activity, then designs such as double sampling and ranked-set sampling discussed in Section 7.4 below could have been evaluated to see if the number of samples (and hence, cost) could have been reduced. These types of evaluations usually require a statistician or someone knowledgeable about environmental sampling designs.

7.3.7.3 For Each Data Collection Design Strategy Determine the Appropriate Number of Samples That Will Satisfy the DQO Outputs. The only design considered for the RMA-4 site was a triangular grid. Hence, the number of samples was only determined for that design.

The number of samples, and hence, the spacing between the intersection points of the triangular grid, depends on the type of statistical test that will be used to compare the RMA-4 and background data. The nonparametric Wilcoxon Rank Sum (WRS) test was selected for RMA-4. This test is recommended in USNRC (1997, page 2-27) to compare radionuclide concentrations with background when the decision unit has a relatively uniform distribution of contamination. [If the radionuclide is not in background, then USNRC (1997) recommends the Sign test.]

The approximate number of samples required from RMA-4 and the background area combined to make the release decision when the WRS test is used is computed using Equation 5-1 in USNRC (1997, page 5-28). This number is then increased by 20% to be conservative. Equation 5-1 requires knowing α and β (tolerable decision error rates) and a parameter denoted by P_r. The values of α and β were specified in Step 6 above (Section 7.3.6.4) to be 0.05 and 0.10, respectively. The parameter P_r is defined to be the probability that a measurement at a randomly selected location in RMA-4 is greater than a measurement at a randomly selected location in the background area. The value of P_r that is specified denotes the extent that the probability distribution of ^{232}Th concentrations at the RMA-4 site overlaps the probability distribution of ^{232}Th in the background area. Hence, the value of P_r specified denotes the amount of overlap in the two distributions that is considered important to detect with probability 1 - β. Allowed values of P_r are between 0.5 and 1.0. If P_r is specified to be 0.50, then one distribution "sits on top of the other one" and the two distributions are identical. If P_r is specified to be equal to 1.0, then the distribution of ^{232}Th at the RMA-4 site is shifted so far to the right (to higher concentrations) of the distribution of ^{232}Th in the background area that there is no overlap whatsoever. If P_r is specified to be a value between 0.5 and 1.0, then the two distributions overlap to some degree. As P_r is specified to be closer and closer to 0.5, then more and more samples will be required to detect the smaller and smaller differences in the two distributions.

In practice, the value of P_r to use is obtained from Table 5.1 in USNRC (1997, page 5-28) or Table 1 in Abelquist (1997, page 9) by first specifying the value of the relative shift, Δ/σ, where Δ is the specified width of the gray region (specified in Step 6 to be 5.92×10^{-3} Bq g^{-1} above background) and σ is the standard deviation among ^{232}Th measurements collected from the RMA-4 area. Abelquist (1997, page 3) used historical soil ^{232}Th measurements for the RMA-4 site to obtain 0.01 Bq g^{-1} (0.27 pCi g^{-1}) as an estimate of σ. Hence, Δ/σ = (5.92×10^{-3} Bq g^{-1}) / 0.01Bq g^{-1} = 0.592, which was rounded to 0.60. From Table 5.1, P_r = 0.66 when Δ/σ = 0.60. (It was assumed that the standard deviation of background area measurements would be no larger than that for the RMA-4 site. To be conservative, the standard deviation for background data was set equal to that for the RMA-4 site (0.01 Bq g^{-1}). This assumption should be checked after the data are obtained. If it is not true, then the larger of the variances of ^{232}Th at the RMA-4 and background sites should be used.)

Using Equation 5.1 in USNRC (1997, page 5-28), the number of samples required in RMA-4 and the background site combined was computed by Abelquist (1997) to be 106.2. Increasing this value by 20% yields 128, which means that 64 soil samples are required in RMA-4 and also in the background area. Then a triangular grid pattern of 64 locations was placed at a random starting point in RMA-4 and also in the background area.

Suppose costs or other practical constraints had indicated 64 samples per area was not feasible. Then various combinations of larger values of α, β, and Δ could have been used in Equation 5.1 in USNRC (1997) to obtain smaller numbers of required samples. Of course, as the number of samples decreases the distance between sample points on the triangular grid will increase, which will change the action level for detecting areas of elevated concentrations (see Step 6).

7.3.7.4 Document the Operational Details and Theoretical Assumptions of the Selected Design in the QAPP, The Field Sampling Plan, The Sampling and Analysis Plan, Or the Decommissioning Plan. The development of the Quality Assurance Project Plan (QAPP) (USEPA 1998a) is a critically important and required step as discussed in USNRC (1997, pages 2-6, 2-7, 9-3 and 9-4). Also, documenting the theoretical assumptions that underlie the selected design is necessary. These assumptions must be evaluated after the design has been implemented (data obtained) in order to assess if the data collected are of the correct type, quantity and quality. For example, some assumptions that underlie the design used at the Cushing site are:

- The standard deviation of the ^{232}Th data at both the RMA-4 and background areas is 0.01 Bq g^{-1}. If this assumption is found to be incorrect, the larger of the two standard deviations computed from the data obtained should be used to re-compute the required number of samples. Additional samples should be collected if the re-computed number is larger than the original number of samples.
- The values of P_r obtained from Table 5.1 in USNRC (1997, page 5-28) are, strictly speaking, only correct if the site and background data are normally distributed. If the collected data indicate the data sets have a large skewness (non-symmetry), then the number of samples collected may be too small to achieve the decision error rates (α and β) that were specified in Step 6 of the DQO process. This problem may be mitigated in practice by the fact that the computed number of samples is always increased by 20%.

- There are no incorrect data (mistakes) in the data sets. Quality checks must be in place to identify mistakes. Statistical tests for outliers can be performed (USEPA 1998b) to identify extreme data values that may, upon further investigation, be found to be mistakes.
- There are no strong spatial correlations among the ^{232}Th data collected on the triangular grid. If this assumption is not true, then the number of samples may be too small. This problem may be mitigated in practice by the fact that the computed number of samples is always increased by 20%.
- There are no non-detects present in the RMA-4 or background data. If all the data have one detection limit and if less than 40% of the measurements in one or both data sets are below this limit, then the WRS test can still be computed using the modified version in Gilbert (1997, pages 247-250). If less than about 15% of the data set are nondetects with either the same or different detection limits, then one could consider replacing each non-detect with one-half the detection limit for that measurement before conducting the WRS test. For all other situations, a statistician should be consulted for guidance.

7.4 SAMPLING DESIGNS

7.4.1 Introduction

The term "sampling design" as used here refers to where and when samples will be collected. Table 7.3 provides a summary of the distinguishing characteristics and the advantages and disadvantages of the following designs: judgment sampling, simple random sampling, systematic (grid or line) sampling, stratified random sampling, double sampling, ranked-set sampling and composite sampling. Section 7.5 considers methods for determining the number of samples that should be used in these designs. Further information is available in Gilbert (1997), Thompson (1992) and USEPA (1999). Also, the U.S. EPA is currently preparing guidance (EPA QA/G-5S) on sampling designs to support quality assurance project plans.

USNRC (1997, page 5-35) and USNRC (1998, page 3-9) recommend systematic sampling using triangular or square sampling grids for final status soil and building surveys at nuclear facilities when the potential for residual contamination after remediation activities is high or moderate (survey units determined to be "Class 1" and "Class 2" units). If the potential of residual contamination is very low ("Class 3" survey units), then USNRC (1997, 1998) recommend simple random sampling since areas of elevated activity are not expected. Also, judgement may be used in Class 3 survey units to determine where radiation scanning of soil should occur. Designs other than systematic and simple random sampling are discussed here because flexibility in applying MARSSIM guidance is possible (USNRC 1997, pages 2-33 through 2-40).

7.4.2 Judgment Sampling

Judgment sampling occurs when an individual or group of individuals use their knowledge of site operations the site conceptual model and past radionuclide surveys and sampling to decide which specific locations or times should be sampled or measured to obtain a data set. As noted above, USNRC (1997) permits the use of judgment sampling for final status surveys in certain specific situations. But care must be taken to prevent its use in situations where the use of personal judgment in selecting sampling locations would seriously compromise the integrity and reliability of influences based on the data.

As noted in USNRC (1997), the use of professional judgement developed on the basis of site operations and prior representative data are critically important in deciding which survey units of a nuclear facility should be classified as Class 1, Class 2, or Class 3 in a final status survey. However, the use of judgment to pin point the specific location of sampling or measuring locations should not be used except when the objective is limited to providing preliminary information about the presence or absence, type, location, or general magnitude of contamination, as stated in Table 7.3.

7.4.3 Simple Random Sampling (SRS) (Table 7.3, Fig. 7.1)

SRS is the simplest probability-based sampling plan for selecting n representative units from the population of interest. Each potential sampling or measurement location or time has an equal probability of being selected. SRS can be used when the goal is to estimate population parameters (mean and percentiles of the population) or to test hypotheses about those parameters. However, other sampling plans discussed below (e.g., systematic or stratified random sampling) are more "efficient," i.e., will provide better answers, if there are trends or spatial patterns over the defined population (geographical or time area) of interest. Also, better answers can often be obtained by more complicated designs (e.g., double sampling and ranked set sampling) if auxiliary information about the likely magnitude of the concentration at potential sampling locations is available. Examples of auxiliary information are quick and inexpensive in-situ measurements of the unit (used in double sampling plans or ranked set sampling) or expert judgment (can be used instead of in-situ measurements in ranked set sampling).

Refer to Table 7.3

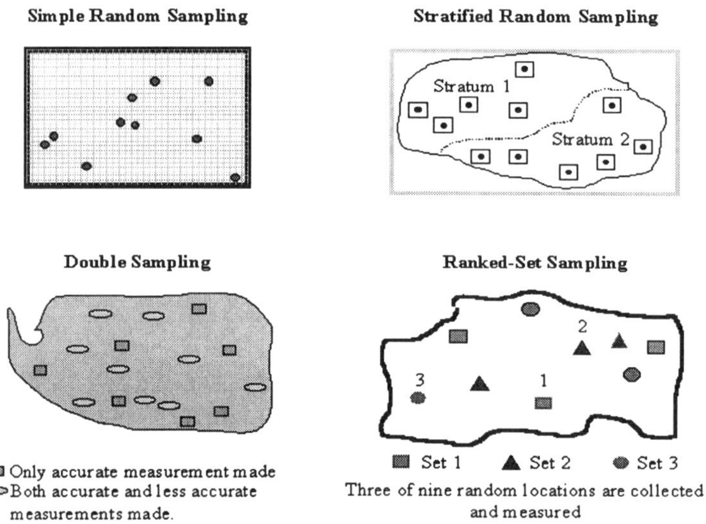

Fig. 7.1 – Four Sampling designs: simple random sampling, stratified random sampling, double sampling and ranked set sampling.

In practice, a set of n random field locations determined using SRS is very easily achieved using the Visual Sample Plan (VSP) software package (see Section 7.6 below). VSP determines the locations, shows them on a map of the site, and provides the geographical coordinates of each location. If VSP is not used, one can determine a random location using a random number table, a calculator or a personal computer to generate random E-W and N-S coordinates. These two coordinates define the random spatial location to be sampled. This process is repeated for each of n locations.

7.4.4 Systematic Sampling (Table 7.3, Fig 7.2)

Systematic sampling consists of taking measurements at locations and/or times according to a spatial or temporal pattern. Examples are collecting units on a square or triangular grid pattern or along a line. Systematic sampling is usually easier to implement than simple random sampling under field conditions because only one random point (one corner of the grid) needs to be located in the field. Systematic sampling is well suited to the goal of estimating the spatial contamination pattern or to look for areas of especially high contamination (hot spots). See Table 7.3 for additional information.

7.4.5 Stratified Sampling (Table 7.3, Fig. 7.1)

Stratified sampling is a design wherein the site is divided into two or more non-overlapping strata (subareas) and units are selected independently within each stratum. If a random start systematic grid is used in each stratum, then the design is called stratified systematic sampling. If SRS is used within each stratum, then the design is called stratified random sampling. Stratified sampling is better than SRS at estimating population parameters and making inferences if the variability among the units within strata is smaller than the variability of the entire population. That is, the units within strata should be more similar to each other than to units in other strata. The strata boundaries are determined using prior knowledge and professional judgment about the patterns of contamination expected in the population.

In addition to providing the potential for more precise answers about parameters such as the mean for the entire site (all strata combined) than can be achieved using SRS, stratified sampling also provides separate estimates of population parameters for each stratum. The number of samples in each stratum may be different, which permits increased sampling in strata of particular interest or strata for which measurements are repeated to be more variable or expensive to obtain.. Measurements obtained using a stratified sampling design should not be used in standard statistical tests of hypotheses or more complicated statistical algorithms without careful consideration. This caution is required because most statistical analysis methods assume that measurements have been obtained using SRS.

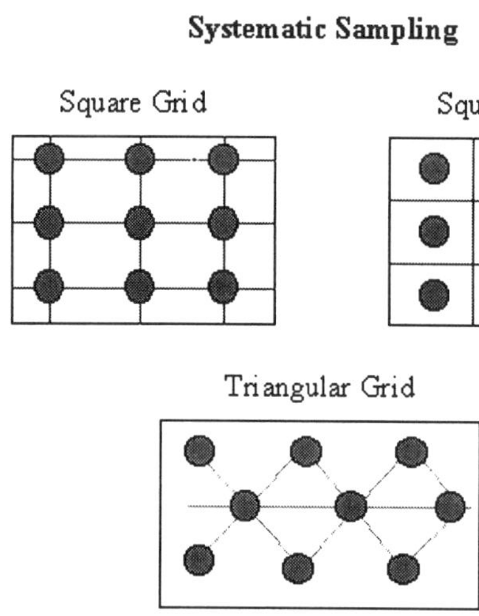

Fig 7.2 – Two types os systematic grid sampling: square and triangular.

7.4.6 Double Sampling (DS) (Table 7.3, Fig. 7.1)

DS can be applied when there are two methods for measuring the amount of a pollutant in an environmental sample, where one of the methods is "accurate" and perhaps expensive and the other is less accurate but also less expensive. Under certain conditions, DS will estimate the mean of the population with greater precision and less cost than can be achieved using SRS. To apply DS, first use SRS to locate a relatively large number of locations in the field at which the "less accurate" measurements will be made. Then, a random subset of these locations will also be measured using the "accurate" measurement method. The data for locations for which both measurement methods were made are used to estimate any linear (straight-line) relationship between the two methods. This fitted linear line is then used to compute the mean concentration for the site and the standard deviation of that estimated mean.

DS should be used only if there is a strong linear relationship (correlation) between the two measurement methods and if the ratio of costs of the accurate to less accurate measurement method is sufficiently large. If developed and applied appropriately, double sampling can give more precise results than by using only the "less accurate" measurements or the "accurate" measurements for a fixed cost. Gilbert (1997, pages 106-118) shows how to determine how high the correlation should be for DS to be useful. He also shows how to determine the number of measurements of each type required. Note that USNRC (1997, pages 4-4 through 4-7) discusses the use of surrogate measurements for determining compliance when multiple contaminants, e.g., ^{137}Cs and ^{90}Sr, are present. When compliance is being evaluated on the basis of the mean ^{90}Sr concentration, DS may be considered as a way of reducing cost by using only a small number of expensive ^{90}Sr measurements in combination with a larger number of less expensive ^{137}Cs measurements.

In concept, there is no reason that DS should be restricted to using only correlation information between the measurement method of choice and *one* type of inexpensive measurement. For example, there may be several types of inexpensive measurements that have been made at the same geographical locations. In that case, one might fit a multiple regression equation rather than just a simple linear regression. However, this author is not aware of any published papers that provide guidance on when to use the multiple regression approach to estimate the mean and on the number of measurements of each type of inexpensive measurement to obtain.

7.4.7 Ranked-Set Sampling (RSS) (Table 7.3, Fig. 7.1)

RSS can result in more precise estimates of means at possibly less cost than if SRS is used. RSS yields samples that better represent the range of concentrations at the site. RSS should be considered for use if there exists a quick and inexpensive qualitative or quantitative method of accurately ordering (ranking) potential sampling locations in the field by the approximate level of the contaminant of concern at those locations. Ranking is done before samples are collected and measured in the laboratory. The quick and inexpensive method might be an inexpensive in-situ field detector measurement or the professional judgment of an expert based, e.g., on his/her visual assessment of potential sampling locations.

RSS may be implemented as follows. To obtain n environmental samples for radionuclide analysis in the laboratory, first locate n x n = n^2 locations (n sets of n locations) in the field using SRS. Rank from lowest to highest (from 1 to n) the *first* set of n locations on the basis of their radionuclide contamination level using the quick, inexpensive method. Similarly rank the *second* set of n random samples from 1 to n, then the *third* set from 1 to n, ..., and finally the nth set from 1 to n. Then only collect and send to the laboratory for measurement the *lowest* ranked location in the *first* set, the *second* highest ranked location from the *second* set, the *third* highest ranked location from the *third* set, ..., and the nth *highest* from the nth group. Then compute the mean using these n laboratory measurements. This average has an expected precision that is smaller than what would be obtained if the average was computed from a SRS of n measurements. The following references provide additional information about RSS: McIntyre (1952), Patil et al., (1994), Gilbert (1995), Patil (1995), USEPA (1995a), Johnson et al. (1996), Muttlak (1996), Samawi and Muttlak (1996), Mode et.al. (1999) and USEPA (1999).

7.4.8 Composite Sampling (Table 7.3, Fig. 7.3)

Composite sampling can, in appropriate situations, more precisely estimate the mean concentration and improve site coverage at less cost than if individual rather than composite samples are used. Also, composite sampling has the potential for improving the power of statistical tests to detect differeences in the means of different populations. Moreover, it can be used in conjunction with other sampling plans, e.g., SRS (Fig. 7.3a) and systematic sampling (Fig. 7.3 b,c,d). In screening surveys, composite sampling can also be used to check for the presence of certain chemical contaminants if the detection limit for the chemical is suitably less than the specified action limit. A disadvantage of composite sampling is that there is a loss of information on individual samples that are pooled to form composite samples. For example, one or more of the individual samples making up the composite sample could exceed a fixed standard. However, if a set of split samples is held in reserve from the original set of individual samples (to possibly be analyzed at a later date), and if the detection limit is suitably less than the action level for the contaminant of concern, then there are ways to efficiently search for those individual samples that are extreme values or "hot spots" (USEPA 1999). Further information on composite sampling is found in Lancaster and Keller-McNulty (1998a, 1998b), USEPA (1995b), Edland and van Belle (1994), and Gilbert (1997).

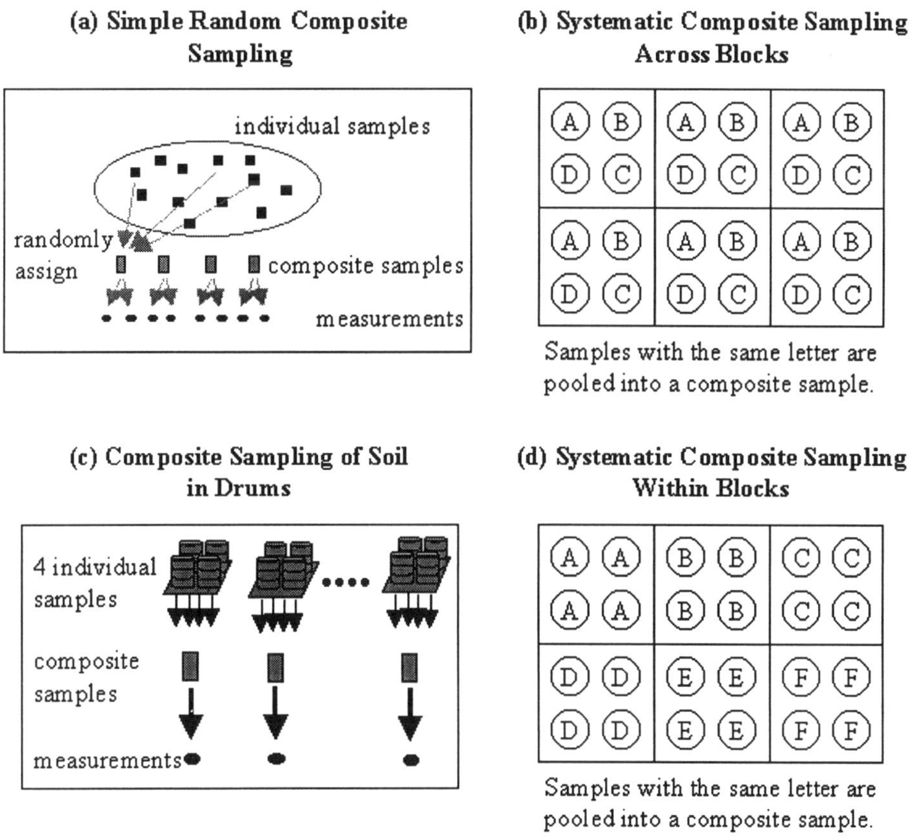

Fig. 7.3 - Four types of composite sampling designs

Table 7.3 Guidance in selecting a sampling strategy

JUDGMENT SAMPLING

Distinguishing Characteristics

- An expert uses expert opinion to decide which locations or times should be sampled or measured

Advantages

- Useful for obtaining preliminary information about the presence or absence, type, location, or general magnitude of contamination based on site operating history and other information

Disadvantages

- The sampling plan is not probability based, i.e., all potential samples do not have a known probability of being selected by the expert. This situation casts serious doubt on any inferences that are made to the underlying population
- The credibility and use of data obtained depends critically on the ability of the expert to pick representative sampling locations and times
- It is not possible to accurately estimate the variance of estimated means, percentiles, or other population parameters
- Data cannot be used to demonstrate compliance with regulatory standards using statistical tests or confidence limits
- Not suitable for making final remedial action decisions based on a final status survey

---

SIMPLE RANDOM SAMPLING

Distinguishing Characteristics

- All possible sets of n sample locations in the target population have an equal chance of being selected. The term "random" does not mean "haphazard."

Conditions for Use

- When no major spatial concentration patterns or areas of elevated concentrations are expected and a uniform coverage of sampling points over the site (such as would be achieved using a grid design) is not required

Advantages

- Provides data appropriate for estimating population parameters (e.g., mean, percentiles) and their variances and for conducting tests of hypotheses
- Can be used in more complex designs such as stratified random sampling, double sampling and ranked-set sampling, or in combination with systematic sampling

- Can be used to form composite samples when estimating means

Disadvantages

- Large gaps between data point locations or times can occur, particularly for large sites.
- Not as good as systematic sampling for estimating trends and spatial patterns
- May be somewhat more difficult to implement in the field than systematic sampling

STRATIFIED SAMPLING

Distinguishing Characteristics

- A heterogeneous site is divided into two or more non-overlapping strata (groups) and samples are collected independently at locations within each stratum using a suitable design such as simple random sampling, systematic grid sampling, double sampling or ranked-set sampling
- The objective is to estimate or make inferences using the mean or total for the *entire* site (all strata combined). Special methods are used to determine the number of samples in each stratum and to estimate the site mean and total based on stratum means and totals (see Gilbert 1997, pages 45-57).
- Note: If the objective is to estimate or make separate estimations or inferences for each separate stratum, then separate sampling designs may be developed for each stratum

Conditions for Use

- When the population has strata with distinctly different average concentrations or degrees of variability but a single decision or estimation for the entire site as a whole is needed

Advantages:

- Provides more representative data for the entire population than if the site is not first stratified into groups
- Provides potential for estimating means and totals for the entire population with greater precision compared to simple random sampling
- Also provides data for obtaining estimates for each separate stratum
- Is generally more cost effective than simple random sampling even if imperfect information is used to construct the strata

Disadvantages:

- Requires prior knowledge of the population to construct strata
- Formulas for estimating the number of samples, mean, total and variance are

more complex than for simple random sampling

- Slightly more difficult to implement in the field than simple random sampling or systematic (grid) sampling

SYSTEMATIC GRID SAMPLING

Distinguishing Characteristics

- Provides for uniform coverage (equal spacing of samples) of the population over space or time starting at a randomly selected location or time
- Can be used to sample over space or along a line

Advantages

- Best sampling plan for estimating boundaries, trends and patterns
- Best for maximizing probability of detecting unknown areas of elevated activity
- Eliminates possibility of having unacceptably large unsampled portions of the population that can arise with sampling plans based on random sampling
- Best sampling plan for estimating the spatial correlation structure (needed for geostatistical applications)
- Possibility exists for use in more complex designs such as stratified sampling, double sampling and ranked-set sampling
- Simpler to implement in the field than other sampling designs

Disadvantages

- Data may yield biased estimates of means and percentiles and incorrect inferences based on statistical tests if the sampling grid pattern happens to line up with a pattern of contamination
- Standard statistical methods to estimate the variance of means and percentiles or to test hypotheses that are appropriate for simple random sampling do not necessarily apply to data obtained using systematic sampling if there are correlations, trends and patterns of concentrations in the field. May need to consult a statistician. See Gilbert (1997, pages 89-105)

DOUBLE SAMPLING

Distinguishing Characteristics

- Uses simple random sampling
- Developed for estimating means and total amounts
- Two types of measurements are used: (1) the preferred or required type of measurement (e.g., alpha spectrometry for $^{239+240}Pu$) and (2) a less expensive but also less precise quantitative auxiliary type of measurement (e.g., gamma scans for ^{241}Am), that is highly correlated with the required measurement type.

- The auxiliary measurements are made for each of a relatively large number of samples, locations or times. The preferred measurements are obtained for a subset of those samples, locations or times.

Advantages

- Can provide more precise estimates of means and totals at less cost than if only the more expensive type of measurement is used
- In many cases systematic sampling is better than simple random sampling to obtain estimates of spatial patterns and trends using the auxiliary measurements

Disadvantages

- Need good prior information on the correlation between the preferred and auxiliary measurements and on the measurement and field costs to evaluate if double sampling is better than other sampling designs.
- Formulas for computing the number of samples, means, totals and variances are more complex than for simple random sampling.

RANKED-SET SAMPLING

Distinguishing Characteristics

- Uses expert opinion or a quantitative inexpensive auxiliary measurement to decide which sampling locations identified using simple random sampling should be sampled for the required measurement method
- Similar in concept to double sampling except that either expert opinion or an auxiliary measurement can be used

Advantages

- Makes use of professional expert opinion in an acceptable way to select sampling locations or times
- On average, will yield more precise estimates of means, totals and percentiles than if simple random sampling is used without expert opinion or auxiliary measurements to select sampling locations or times
- Has been shown in certain cases to increase the ability of the Wilcoxon Rank Sum (WRS) test to detect differences between two populations (Bohn and Wolfe 1992, 1994)
- Has been shown in certain cases to increase the ability of the Sign test to detect when the median concentration exceeds an action level (Koti and Babu 1996)
- Performs better than double sampling for estimating means and totals unless the correlation between the auxiliary measurement and the required measurement is very high (> 0.85)

Disadvantages

- Must use simple random sampling to locate a total of n^2 locations in the field in order to select n that are actually sampled
- Must carefully evaluate field and measurement costs to determine if ranked-set sampling is preferred
- The performance advantage of ranked-set sampling requires the expert to accurately order (rank) from smallest to largest sampling locations determined using simple random sampling with regard to their radionuclide concentration
- Use of data obtained using ranked-set sampling to test hypotheses is currently only available for the WRS test and the Sign test (Hettmansperger 1995; Koti and Babu 1996; Barabesi 1998), although these tests are the ones recommended by USNRC (1997)

COMPOSITE SAMPLING

Distinguishing characteristics

- Sets of individual samples are pooled together and thoroughly mixed to form a composite samples
- Ideally, the composite sample is a perfect physical averaging of the individual samples

Advantages

- The precision of estimated means is increased if analytical and measurement variability is small compared to variability among individual samples
- Can reduce the cost of estimating the mean by reducing the number of measurements made
- Can be used to help locate areas of elevated concentrations

Disadvantages

- Regulators may not permit the use of composite sampling
- Composite samples cannot be used if decisions must be made using the sample support of individual samples rather than that of composite samples
- Costs may be higher than for non-compositing designs if costs of creating composite samples are high relative to analytical costs.
- The sample matrix may impede thorough mixing of the individual samples
- Must guard against a loss of radionuclides during the mixing process

7.5 NUMBER OF SAMPLES

Developing a design requires determining the number of samples that should be placed over space or time. The number of samples depends on several factors, as pointed out in Sections 7.3.6 and 7.3.7 (Steps 6 and 7 of the DQO process). A key factor is the particular statistical test that will be used. USNRC (1997) recommends using the Wilcoxon Rank Sum (WRS) test to compare the site median to the background median and the Sign test to compare the site median to a fixed threshold value. The number of samples required for these two tests when simple random sampling or systematic sampling is used is computed by the VSP software code discussed in Section 7.6 below. The equation used by VSP for the WRS test is the same as that given in USNRC (1997, page 5-28). Similarly, the equation used by VSP for the Sign test is the same as given in USNRC (1997, page 5-33). However, note that VSP does not add on to these computed sample sizes the additional 20% recommended by USNRC (1997, pages 5-29, 5-33).

VSP also computes the number of samples required for the one-sample t test (for comparing the mean of a normal distribution to an action level), the two-sample t test (for comparing the means of two normal distributions, e.g., a site mean versus a background mean), a confidence interval for a mean, the standard deviation of a normal distribution, the nonparametric Wilcoxon Signed Rank test to compare a median of any symmetric distribution to an action level, nonparametrically comparing an estimated proportion to a specified proportion, and nonparametrically comparing two proportions. VSP requires that the user provide values for key parameters that are outputs of Steps 5 and 6 of the DQO process, e.g., tolerable Type I and Type II decision error rates (α and β), the anticipated variance of the data, and the width of the gray region.

7.6 VISUAL SAMPLE PLAN (VSP) SOFTWARE

7.6.1 Introduction

Visual Sample Plan (VSP) is a computer software program that answers two important questions based on information obtained by going through the DQO process.

1. How many samples are needed?
2. Where should the samples be taken?

VSP can quickly calculate the number of samples needed for various scenarios at different costs. It also instantly provides random or gridded sampling locations overlaid on the site map.

The purpose of VSP is to provide simple, defensible tools for selecting and visually displaying a sampling design and number of samples using inputs obtained from the DQO process. VSP can be used for any two-dimensional sampling plan including surface soil, building surfaces, water bodies, or other similar applications. VSP includes a odule called SampTOOL, which is an interactive question-and-answer flow diagram that helps the user determine which sampling strategy(ies) should be considered for application. The latest beta version of VSP can be downloaded from the following web site: http://terrassa.pnl.gov:2080/DQO/software/vsp. VSP development has been partially supported by the Department of Energy (DOE) National Analytical Management Program and is presently a joint effort between Pacific Northwest National Laboratory and Oak Ridge National Laboratory - Grand Junction

7.6.2 Features of VSP

- Uses statistical and mathematical algorithms applicable to environmental statistics and MARSSIM
- Interacts with the user through familiar visual interfaces such as site maps and building plans
- Provides immediate feedback of the projected results of selected statistical sampling plans by overlaying random sampling locations or grids directly onto the site map or building plan
- Provides projected number of samples, total sampling costs, and sampling locations in appropriate coordinates
- Includes an option to use *quasi-random sampling* to obtain sampling locations that are randomly but more evenly (with smaller spatial data gaps) spaced than what typically occurs with simple random sampling
- Includes *adaptive-fill sampling*, which determines sampling locations to fill in spatial data gaps in a set of sampling locations previously determined using authoritative (judgment) sampling, simple random sampling, or systematic sampling

- Provides graphic decision tools such as graphs of probability of hot spot detection vs. total sampling costs. The aim is to make all output as simple, clear, and visually appealing as possible
- Allows for determining the number of samples needed to achieve the DQOs specified for either nonparametric statistical test (like those in USNRC 1997) or parametric statistical tests
- Incorporates SampTOOL, a tool to guide the user to the appropriate type of sampling design (Section 7.6.3.6)

7.6.3 Quick-Start Guide to VSP

This section introduces the basic features of VSP the program. To learn most efficiently from this example, the user should have VSP downloaded and in use on their computer.

7.6.3.1 ***Open the Project.*** Open the project named "Example1.vsp". The project may be opened by selecting **File / Open Project** on the menu and choosing *Example1.vsp* in the **File Open** dialog box. The File Open dialog box may also be accessed by clicking the **Open** button of the toolbar or by using the keyboard shortcut **Ctrl-O**.

7.6.3.2 ***Map View.***

- When first opened, the **Map view** is the default view for the project. The Map view shows a plan view of the project site. The area colored red is the survey unit, or "sample area." Samples only appear inside of sample areas. In this project, the samples are the light blue circles with crosses in them. A project can contain many sample areas. Expand the project window by pressing the **Maximize** button on the upper right corner of the project window.
- On the Map view, you can zoom in to examine details of the project site. Choose **View / Zoom In** from the menu (or click on the **Zoom In** button on the toolbar, which is the magnifying glass with a plus sign in it). The cursor becomes a magnifying glass. Clicking on the map with the magnifying glass will center that point in the window and make the view larger. Click a couple of times on the map to see how it is magnified. (Note: there are built-in limits on how much the view can be magnified. Once the limit is hit, the zoom mode will turn off.)
- You can also zoom out to see a larger area of the project map. Choose **View / Zoom Out** from the menu (or click on the **Zoom Out** button on the toolbar,

which is the magnifying glass with the minus sign in it).). The cursor becomes a magnifying glass. Clicking on the map with the magnifying glass will center that point in the window and make the view smaller. Click a couple of times on the map to see how it shrinks. (Note: there are built-in limits on how much the view can be shrunk. Once the limit is hit, the zoom mode will turn off.)

- To see the site map as large as possible while keeping it all visible in the window choose **View / Zoom Max** from the menu. Do this now to restore the map view.
- To focus in on a specific area of the site map use the Zoom Window feature. Choose **View / Zoom Window** from the menu (or click on the **Zoom Window** button on the toolbar). The cursor becomes a magnifying glass with a cross inside it. Take the cursor to the upper left corner of the sample area. Press *and hold* the left mouse button while dragging the cursor to the lower right corner of the sample area. (You will see a rectangle being drawn on the map – this indicates the area that will be magnified.) When the rectangle encloses the sample area release the mouse button and the map view will be magnified to show the area selected. If you didn't quite get the area you wanted, do **Zoom Max** and try again. (Note: the rectangle selected may not end up in the center of the window if the built-in zoom limit is reached.)
- The sample area can be turned on or off (**selected / deselected**). Put the cursor on the sample area and left-click the mouse. The sample area is now *deselected* and is shown as a colored outline. A deselected sample area is protected from sampling operations. This is important when there are several sample areas on a map, and you want to perform different operations on them. Click on the sample area again to select it. (These operations can also be accomplished by using **Edit / Sample Areas / Select/Deselect Sample Areas** on the menu.)
- Information about individual sample points is available by right clicking on them with the mouse. Try this and note the dialog box that pops up. It shows the sample type, coordinates, and label. The label is empty by default, but can be changed by the user. Enter some text in the Label box and press the OK button. Right click on the sample again and note that the label has been retained.

7.6.3.3 Graph View.

- The **Graph view** shows a graph relating to some aspect of the sampling design. Certain sampling designs do not have an associated graph. Select the graph view by clicking the **Graph View** button on the toolbar. (It may also be selected using **View / Graph** on the menu.)

- This sampling plan used a one-sample proportion test. The associated graph shows the beta error rate vs. number of samples. The graph is interactive. You can move the cursor to a value on the x-axis (beta error rate in this case) and see the related value on the y-axis (number of samples in this case). These cross hairs can be turned on or off by choosing **Options / Graph / Display Cross Hairs** from the menu. The values also appear on the status bar at the bottom of the window in numerical format
- Graphs also show the current value of the parameter on the x-axis. It appears as the dashed blue line. The current value is the value of the parameter used for the current sampling design. A beta error rate of 20% was used for this sampling design. The blue line can be turned on or off by choosing **Options / Graph / Display Current Value** from the menu.
- If one of the axes is *Number of Samples* it can also be displayed as *Sampling Cost*. To change the axis, choose **Options / Graph / Display Cost**.

7.6.3.4 ***Report View.*** The **Report view** shows text about the current sampling design. It shows: the type of sampling (random, grid, hotspot), size of grid (for grid sampling), number of samples, total cost, number of sample areas, total area of sample areas, and probability of hit (for hotspot sampling). Select the report view by clicking the **Report View** button on the toolbar. (It may also be selected using **View / Report** on the menu.)

7.6.3.5 ***Coordinate View.*** The Coordinate view shows the coordinates for each sample point on the map. Coordinates are segregated by sample area. These coordinates can be copied and pasted into a spreadsheet or word processor.

7.6.3.6 ***SampTOOL.*** SampTOOL is a web-based tool that leads you through a series of questions to assist you in selecting the type of sampling design to best suit your sampling needs. To access SampTOOL, choose **Advisor / SampTOOL** on the menu. The dialog box allows you to choose your preferred browser to view SampTOOL. On the radio buttons choose the browser (Netscape, Explorer, or Other). Press the **Launch** button to view SampTOOL. If you receive an error message that says "Invalid Path" you need to correctly specify the path to your browser. You can use the browse button to assist in finding the path. Please refer to SampTOOL documentation for further information about its features and usage.

7.6.3.7 Miscellaneous.

- **Edit / Copy** can be used to transfer information to other Windows programs. The map view and graph view can be placed in the clipboard and then be placed in a graphics program such as Windows Paint. The report view text can be transferred to a word processor. The coordinate view can be transferred to a word processor or spreadsheet.
- **Edit / Paste** can be used to transfer sample information into a sampling project. Paste is only active on the coordinate view. The information must be text from a spreadsheet, text editor, or word processor. It must contain at a minimum the X and Y coordinates of each sample point to be placed in the project. The coordinates must fall inside a sample area or they will be ignored. Additionally, the information may contain a sample type and a label.
- **Largest Unsampled spot** can be used to see where deficiencies in sample coverage may exist. It shows where the largest circle can be placed inside each sample area without overlapping a sample. To see how this works, choose **Tools / Largest Unsampled Spot / Find**. A dialog box appears. Accept the default parameters by pressing the OK button. It may take a few moments to find, depending on the number of samples, the complexity of the sample area, the accuracy requested, and the speed of your computer. When the spot is found, it will be display along with a dialog box describing its size. The spot can be hidden by choosing **View / Largest Unsampled Spots** on the menu or by clicking the **Show Unsampled Spots** button on the toolbar.
- **Double / Triple / Quad Window**. Multiple windows can be opened for a single project, each with a different view. The easiest way to achieve this is by choosing **Window / Quad Window** from the menu. This opens four windows, selects the four views (map, graph, report, coordinate), and arranges them on the screen. **Double Window** shows only the map and graph views. **Triple Window** shows the map, graph, and report views.

7.7 REFERENCES

Abelquist, E.W. Final status survey using MARSSIM survey methodologies at the Cushing refinery site. Oak Ridge Institute for Science and Education, Oak Ridge, TN, July 1997.

Barabesi, L. The computation of the distribution of the sign test statistic for ranked-set sampling. Communication in Statistics-Simulation 27(3):833-842; 1998.

Bohn, L.L. and Wolfe, D.A. Nonparametric two-sample procedures for ranked-set samples. J. of Amer. Statistical Assoc. 87:552-561; 1992

Bohn, L.L. and Wolfe, D.A. The effect of imperfect judgment rankings on properties of procedures based on the ranked-set samples analogue of the Mann-Whitney-Wilcoxon statistics. J. of Amer. Statistical Assoc. 89:168-176; 1994.

Edland, S.D. and van Belle, G. Decreased sampling costs and improved accuracy with composite sampling. In: Cothern, C.R.; Ross, N.P., eds. Environmental statistics, assessment, and forecasting, Lewis Publishers; 1994:29-55.

Gilbert, R.O. Statistical Methods for Environmental Pollution Monitoring, New York, NY: Van Nostrand Reinhold; 1987. Now available from Wiley & Sons, New York, NY, 1997.

Gilbert, R.O. Ranked set sampling, In: DQO statistics bulletin: statistical methods for the data quality objectives process 1:1-12, Pacific Northwest Laboratory, Richland, WA, PNL-SA-26377; 1995.

Hettmansperger, T.P. The ranked-set sample sign test. Nonparametric Statistics 4:263-270; 1995

International Atomic Energy Agency (IAEA). Compliance Monitoring for Remediated Sites. INIS Clearinghouse, International Atomic Energy Agency. IAEA-TECDOC-1118, Vienna, Austria, October 1999.

Johnson, G.D. Nussbaum, B.D., Patil, G.P. and Ross, N.P. Designing cost-effective environmental sampling using concomitant information, Chance 9:4-11; 1996.

Koti, K.M. and Babu, G.J. Sign test for ranked-set sampling. Communication in Statistics-Theory and Methods 25(7):1617-1630; 1996.

Lancaster, V. and Keller-McNulty, S. Composite sampling, part I. Environmental testing & analysis, July/August 1998a.

Lancaster, V. and Keller-McNulty, S. Composite sampling, part II. Environmental testing & analysis, September/October 1998b.

McIntyre, G.A. A method of unbiased selective sampling, using ranked sets, Australian J. of Agricultural Research 3:385-390; 1952.

Mode, N.A., Conquest, L.L., and Marker, D.A. Ranked set sampling for ecological research: accounting for the total costs of sampling, Environmetrics 10:179-194; 1999.

Muttlak, H.A. Pair rank set sampling. Biometrics Journal 38:879-885; 1996.

Patil, G.P., Editorial: Ranked set sampling, Environmental and Ecological Statistics 2:271-285; 1995.

Patil, G.P., Sinha, A.K. and Taillie, C. Ranked set sampling for multiple characteristics. International Journal of Ecology and Environmental Sciences 20:357-373; 1994.

Samawi, H.M. and Muttlak, H.A. Estimation of ratio using rank set sampling. Biometics Journal 38:753-764; 1996.

Thompson, S.K. Sampling. Wiley & Sons, New York, NY, 1992.

U.S. Environmental Protection Agency. Data Quality Objectives Process for Superfund. Interim Final Guidance. Office of Solid Waste and Emergency Response, Washington, D.C. EPA 540-R-93-071, September 1993.

U.S. Environmental Protection Agency. EPA observational economy series, volume 2: ranked set sampling, Office of Policy, Planning, and Evaluation, EPA-230-R-95-006, August 1995a.

U.S. Environmental Protection Agency, EPA observational economy series, volume 1: composite sampling, Office of Policy, Planning, and Evaluation, EPA-230-R-95-005, August 1995b.

U.S. Environmental Protection Agency. Guidance for the Data Quality Objectives Process. EPA QA/G-4, Office of Research and Development, Washington, D.C., EPA/600/R-96/055, September 1994.

U.S. Environmental Protection Agency. EPA Guidance for Quality Assurance Project Plans. Quality Assurance Division, Washington, D.C. EPA QA/G-5, EPA/600/R-98/018, 1998a.

U.S. Environmental Protection Agency. Guidance for the Data Quality Assessment: Practical Methods for Data Analysis. EPA QA/G-9, QA97 Version, EPA/600/R-96/084, 1998b.

U.S. Environmental Protection Agency. RCRA Waste Sampling Draft Technical Guidance, SW-846 Chapter Nine, Planning, Implementation, and Assessment. Solid Waste and Emergency Response, Washington, D.C., EPA 530-R-99-015, July 1999.

U.S. Nuclear Regulatory Commission. Multi-Agency Radiation Survey and Site Investigation Manual (MARSSIM). Springfield, VA: The National Technical Information Service; NUREG-1575 (EPA 402-R-97-016), December 1997.

U.S. Nuclear Regulatory Commission. A Nonparametric Statistical Methodology for the Design and Analysis of Final Status Decommissioning Surveys. Springfield, VA: The National Technical Information Service; NUREG-1505, 1998.

7.8 AUTHOR PROFILES

Dr. Richard O. Gilbert obtained his M.S. in Statistics at Kansas State University in 1965 and his Ph.D. in Biomathematics at the University of Washington in 1969. After joining Battelle in 1969, his professional experience has included the development of environmental sampling plans and data analyses, applied research, teaching, facilitating the Data Quality Objectives Process for environmental cleanup studies, and project management. He authored the widely used book Statistical Methods for Environmental Pollution Monitoring, Van Nostrand Reinhold, 1987 (now published by Wiley & Sons), 26 peer-reviewed papers and 60 reports, all applied to statistical aspects of environmental pollution problems. From 1971 to 1986 Dr. Gilbert developed statistical designs and conducted statistical analyses for nuclear weapons testing sites on the Nevada Test Site for the U.S. Department of Energy (DOE), Nevada. From 1975 to 1984, Dr. Gilbert wrote and published for the DOE 27 issues of the periodical TRAN-STAT (Statistics for Environmental Studies). From 1988 to 1993, Dr. Gilbert was Statistics Task leader of the Hanford Environmental Dose Reconstruction Project. From 1989 to 1992 he managed and was co-principal investigator for a U. S. Environmental Protection Agency (EPA) project to write a document (published by EPA in 1994) to provide statistical designs and tests for evaluating the attainment of reference-based cleanup standards at Superfund sites. Since 1992, Dr. Gilbert has facilitated and/or taught the EPA-developed Data Quality Objectives (DQO) planning process and related environmental statistical methods at numerous DOE sites as well as for the U.S. Air Force, Navy and Army. In 1999 Dr. Gilbert co-authored a handbook for the U.S. Navy that describes statistical methods for assessing Navy site versus background concentrations at Navy facilities in California.

Dr. Gilbert is a member of the N13.31 Working Group that is preparing the ANSI standard "Assessment of Radiation Doses from Plutonium and Americium from Soil." In 1995 he was elected Chair of the ASA Section on Statistics and the Environment. He was a member of Scientific Committee 64-17 (Uncertainty in Environmental Transport in the Absence of Site Specific Data) of the National Council on Radiation Protection and Measurements. Also, Dr. Gilbert is a Fellow of the American Statistical Association (ASA) and a consultant to the U.S. EPA Science Advisory Board.

Mr. Wilson spent several years after High School honing his programming skills in assembly language. In 1987, after graduating from Mesa State College in Grand Junction, Colorado with an Associate's degree in computer information systems, he went to work for a small company in Costa Mesa, California writing computer games. Mr. Wilson returned to Grand Junction in 1990 to work for Oak Ridge National Laboratory

(ORNL) as a computer programmer. Since 1992 he has been the computer services manager for the ORNL office in Grand Junction. In 1998 he graduated summa cum laude from Mesa State College with a bachelor's degree in computer information systems. Mr. Wilson continues to write database and technical software applications for the PC Windows environment. Visual Sample Plan (VSP), developed in conjunction with James Davidson of Pacific Northwest National Laboratory, has been one of his highest achievements.

Chapter 8

BAYESIAN THINKING

Tom Tadfor Little

8.1 INTRODUCTION

The process of Bayesian inference, and its application to radiological dose assessment, is detailed in many publications. This section does not attempt to duplicate these technical presentations, but is intended instead to promote an intuitive understanding of the Bayesian approach by presenting simple examples that illustrate the essence of the Bayesian view of probability.

The essential feature of Bayesian statistics is that the quantities of interest are assumed to follow some probabilistic distribution, even before measurements are undertaken to determine their actual value. For example, even before I measure the length of the desk in my office, I know that the probability that the length is less than 1 millimeter or greater than 1 kilometer is extremely small, if not 0. It would be unreasonable for me to interpret a measurement result of 10.3 kilometers with the same credulity as I would interpret a result of 1.03 meters. The measurement result does not exist in a vacuum; rather it exists against a backdrop of expectations. Bayesian methods capture these expectations quantitatively in the form of a prior probability distribution for the measured quantity.

Although one may establish the prior using only objective information (perhaps a set of previous measurements of the table), the Bayesian approach lends itself quite naturally to the incorporation of subjective judgments. The use of subjective probability in Bayesian statistics is a key point of controversy. This section is intended to clarify this issue, as well as illustrating the basic method of Bayesian inference.

8.2 DEFINITIONS OF PROBABILITY

Although we all work with the concept of probability, and may consider it quite familiar and mundane, there are serious philosophical issues involved in defining probability. The three most influential definitions of probability follow.

Classical. The possible outcomes of an event are analyzed into a set of elementary, equally probable events. The probability of a particular outcome is then the number of elementary events producing the outcome divided by the total number of elementary events. For example, the probability of an outcome of 8 on rolling two dice is obtained by counting the number of 2-dice combinations (out of 36 total) that yield 8. The definition presumes we have some *a priori* way to identify equally probable elementary events. This is sometimes possible from symmetry considerations, sometimes not.

Frequentist. The probability of an outcome is its frequency in the limit of an infinite number of trials. A die might be rolled 100, 1000, or 1,000,000 times, for example, and the number of "2" outcomes in each case might be 21, 159, and 166,034. The frequentist estimate of the probability of obtaining "2" would thus be near 0.166, with some uncertainty. This definition presumes it is possible to perform consistent, repeated trials of the event, at least in theory if not in practice.

Subjective. The probability of an event is based on the observer's assessment of a coherent bet. If I would be equally likely to bet for or against the outcome of "2" on the roll of a single die when the odds are set at 5:1, then the probability of obtaining a "2" is 1/6. With the subjective definition, different observers need not agree on the probability of an outcome.

Each of these definitions has strengths and weaknesses.

The classical definition allows probability to be determined without uncertainty or ambiguity, and without needing to appeal to anything outside the problem description. From a philosophical point of view, of course, it is entirely question-begging. On what basis do we presume to know that the elementary events we have identified are equally probable? Even if a die is symmetrical, why does it follow that each face is equally likely to appear on top after a roll? If one answers, "experience tells us so," then one has abandoned the classical definition in favor of a frequentist one.

The frequentist definition is thoroughly empirical, which makes it philosophically comfortable in connection with the experimental sciences. It has the distinct advantage over the classical definition that it does not require a detailed knowledge of the underlying structure of the problem being investigated. It can be used even when the underlying elementary events cannot be identified. It can, however, take advantage of the classical approach when convenient: when the elementary events are well known, the frequentist can use a classical analysis to anticipate the outcome of repeated trials, without having to actually conduct them. Philosophically, however, the frequentist definition raises its own questions: What does it mean to repeat a trial? If all conditions were duplicated exactly, would not the outcome be the same each time? Should one even speak of probability for individual events using the frequentist definition, or only of probability for ensembles of events? In practical terms, the frequentist definition cannot be applied to unique events, such as a presidential election.

The subjective definition has the merit of universal applicability. It can be used to define probabilities for any event whose outcome is not known. It also subsumes the classical and frequentist approaches, because a classical or frequentist argument can certainly form the basis for an observer's subjective belief in the likelihood of a given outcome. The subjective definition, however, also allows for the incorporation of additional information about the problem, even when there is no formal objective process for doing so. This is both its strength and its weakness. It allows all possible considerations to be included, but at the expense of any guarantee that observers will obtain the same answer. I may feel that the person rolling the die is dishonest, and that the probability of "2" is thus much less than 1/6. You may disagree. Even after sharing all our collective information about the person, we may still disagree. Such practical issues aside, though, the subjective definition of probability is unproblematic philosophically. It simply quantifies a particular observer's degree of belief in the likelihood of a given outcome.

Subjective probability, it might be feared, would produce nothing but a free-for-all in which different observers each indulge their prejudices and no reliable, objective conclusions can be expected to emerge. This possibility is mitigated by two important factors.

The first is the idea of the coherent bet. The subjective definition of probability does not allow me to assign any arbitrary number to my degree of belief. I am obliged to find the particular number at which I would expect to "break even" on a bet. Even if I want Al Gore to be elected president, I can't assign a probability of 1 to this outcome, because then I would be obliged to wager an indefinite sum of money against a payback of 0, which would hardly be sane. The idea of the coherent bet enforces self-honesty in the assignment of probabilities, although it certainly does not guarantee objectivity.

The second, and most important, factor that allows subjective probability to yield reliable conclusions is that the usual rules of calculating with probability still apply. In particular, Bayes's Rule defines how the probability of an outcome must change when new data becomes available. It applies regardless of whether the initial probability was obtained through classical, frequentist, or subjective considerations. So although the subjectivist may have great latitude in assigning initial probabilities, the updating of those probabilities in the light on new data is constrained by mathematics, and the subjectivist is obliged to accept its results.

Finally, it is worthwhile to draw attention to the limitations of objectivity in the sciences. We may strive for an ideal situation in which all conclusions follow inevitably from information that all parties have equal confidence in. In practice, though, we often find ourselves in a position of uncertainty regarding factors and possibilities that cannot be subjected to definitive testing. We are like the observers wondering whether or not the dice-roller is honest. In such circumstances, the fact that different observers differ in their degree of belief in the likelihood of a particular outcome is not a shortcoming, it is an asset. It calls to our attention the limitations of the available data. Under such circumstances, a "see no evil" approach of attending too only the rigorously objective information may lead to error; it may be quite unwise for me to ignore my hunch about the die-roller. The use of subjective probability does no harm when the objective information is copious and compelling. But when the objective information is sparse and ambiguous, the use of subjective probability helps us avoid error, by making our preconceptions explicit and forcing us to consider their role in interpreting our observations. Scientific measurements help us to move from subjective ambiguity to objective confidence. Subjective probability provides a consistent framework for tracking that entire process; the classical and frequentist definitions are applicable only in the final stages, when objective confidence has been essentially achieved.

8.3 BAYES'S RULE

Bayes's Rule provides a means for updating probabilities in response to data. For illustration purposes, we will consider the case of n discrete hypotheses, H_i, where i ranges from 0 to n-1. If the initial probability that H_i is true is $P[H_i]$, then the probability that H_i is true after datum d has been observed is given by Bayes's Rule as

$$P[H_i|\, d] = A \times P[d|H_i] \times P[H_i],$$

where A is a normalization factor given by $1/A = \Sigma_i\ (P[d|H_i] \times P[H_i])$. The likelihood function $P[d|H_i]$ is usually easily calculated from a knowledge of the hypothesis and the measurement process that yields the observed data. The updating procedure can be expressed conveniently as an iterative algorithm:

for each hypothesis H_i :
    calculate the likelihood of the observed data, $P[d|H_i]$
    multiply the likelihood by the prior probability of the hypotheses $P[H_i]$
then renormalize the resulting set of products to form the posterior probabilities $P[H_i|\, d]$

The algorithm can implemented as a spreadsheet, which makes an excellent tool for studying the behavior of Bayesian updating in different situations. The examples that follow are intended for use with such a spreadsheet learning tool.

8.3.1 Example 1: Medical Diagnosis

Suppose a certain medical diagnostic test is 95% accurate when given to individuals who have the disease, and 98% accurate when given to those who do not. A statistical study of a random population sample implies that the disease occurs in the population with a frequency of 0.004 (4 cases per thousand individuals). If a person is selected at random from the population and given the diagnostic test, what is the probability that the person tested has the disease, given a positive test result?

Solution. Define the hypotheses as

$$H_0 = \text{subject does not have the disease}$$
$$H_1 = \text{subject does have the disease}$$

The prior probabilities are inferred from the results of the statistical study. That is,

$$P[H_0] = 0.996$$
$$P[H_1] = 0.004$$

The likelihood values are taken from the accuracy data on the test. Let $d = 0$ signify a negative test result and $d = 1$ a positive test result. Then

$$P[0|H_0] = 0.98 \quad P[0|H_1] = 0.05$$
$$P[1|H_0] = 0.02 \quad P[1|H_1] = 0.95$$

Given a $d = 1$ test result, the product of prior and likelihood for H_0 is 0.02 x 0.996 = 0.01992 and the product for H_1 is 0.98 x 0.004 = 0.00392. The sum of these products is 0.02384, so the normalization factor is the reciprocal of this, or 41.946. Hence the final posterior probabilities are

$$P[H_0] = 41.946 \text{ x } 0.01992 = 0.836$$
$$P[H_1] = 41.946 \text{ x } 0.00392 = 0.164$$

Hence the probability that the subject has the disease, given the positive test result, is only 16.4%.

Comments. The solution of this problem assumes either the frequentist or the subjective definition of probability. The classical definition cannot be easily applied, because we do not know what "elementary equally probable events" are involved in a person contracting a disease. The statistical study cited constitutes a frequentist estimate of the prior probabilities. Presumably the figures for the accuracy of the test also depend on such empirical data.

Suppose no such population data were available. Would it be wise to use a "flat prior", $P[H_0] = P[H_1] = 0.5$, in effect assuming that any positive test result is 95% likely to be true, and any negative result 98% likely to be true? This approach may seem "unbiased", but in fact it will give good results only when the frequency of the disease in the population is near 50%. Even if nothing at all is known about the population frequency of the disease in question, "common sense" seems to urge against such an assumption! In such a case, the frequentist cannot legitimately proceed to solve the problem. The subjectivist can, perhaps guessing at a population frequency of 0.001, knowing quite well it may be wrong by an order of magnitude or more.

8.3.2 Example 2: Detection of Radioactivity

Assume that a sample of material is tested for the presence of radioactivity by use of a counting technique. We will simplify grossly and assume that the measurement process reports a result at one of the 11 discrete levels, from 0 to 10 units, and that the measurement uncertainty is represented by a Gaussian function with standard deviation 2.

Suppose the sample is submitted as a part of procedure to test the measurement system. A lot of 100 samples is received, and it is known in advance that 99 samples are blank (true activity = 0) and 1 sample is "spiked" (true activity = 5). One sample is taken at random from the lot and measured. The measurement result is 3. What is the probability that this sample is the spiked one?

Solution. The prior is obtained directly from the information given about the sample lot.

$$P[H_0] = 0.99$$
$$P[H_5] = 0.01$$
$$P[H_i] = 0, \text{ for all other values of } i$$

The likelihood function is simply the Gaussian measurement uncertainty, centered on the measurement result $d = 3$. The point $x = 5$ is 1 standard deviation above the mean; the point $x = 0$ is 1.5 standard deviations below the mean. Hence the relevant likelihood values are

$$P[3|H_0] = 0.0648$$
$$P[3|H_1] = 0.1210$$

Performing the multiplication and renormalization, the posterior probabilities of the two hypotheses are

$$P[H_0] = 0.981$$
$$P[H_5] = 0.019$$

Hence the probability that this sample is the spiked one is only 1.9%, despite the relatively high measurement result.

Comments. The prior in this case is obtained from thoroughly classical considerations; we know there are 100 equally probable elementary events, and we know how many correspond to each of the possible outcomes. As in the case of the medical diagnosis, the high rate of "zeroes" in the population results in a relatively low confidence that radioactivity (or disease) has been positively detected, even when the measurement (test), taken at face value, suggests its presence.

It is instructive to repeat this example with a higher measurement result (4, 5, or 6), as well as with a smaller measurement uncertainty (1 or 0.5).

In cases such as those presented in these examples, the result of applying Bayes's Rule may seem counterintuitive if one is used to taking a measurement result as "gospel." However, the result is an inescapable consequence of the population frequencies (prior probabilities) used in the examples. In neither case does the result depend on subjective probability.

8.3.3 Example 3: Detection of Radioactivity with a Subjective Prior

Assume a situation identical to that in the previous example, except that now the sample in question is a urine bioassay sample from a routine monitoring program. In other words, there were no workplace indicators to suggest that any intake of radioactive material had occurred. The measurement result is 5 (i.e., 2.5 standard deviations above 0). What is the probability that no intake of radioactive material at all has occurred?

Solution. The solution obviously depends on the prior probabilities one assumes for the likelihood of different levels of radioactivity in the subject's urine. All that is known is that the individual is participating in a routine bioassay program, and that there were no workplace indicators suggesting an intake. Perhaps the frequency of intakes of different magnitudes for this "routine population" can be estimated from population data at the facility, perhaps not.

It seems reasonable to choose a prior that is large at $x = 0$ and decreases rapidly as x increases. Literature search or model intake scenarios might suggest a suitable prior. For illustration purposes, we consider two cases, a prior only weakly weighted toward 0 intake, and a prior very strongly weighted toward 0 intake:

(a) With a prior given by $A/(x + 1)$, where A is a normalization factor, the probability that no intake has occurred is 4.5%. This prior assumes that intakes undetected by workplace indicators are relatively common, occurring in about two-thirds of all cases.
(b) With a prior given by $A/(x + 0.01)$, the probability that no intake has occurred is 78.5%. This prior assumes that undetected intakes are quite rare, occurring in less than 3% of cases.

Comments. The choice of prior clearly has a dramatic effect on the final assessment of the probability of intake. It is not something that can be ignored (i. e., assumed to be flat) without significant consequence. Even in the absence of extensive population data, it is likely that the health physicist will be able to judge which of two widely different priors (like those in this example) is more realistic. Even such a broad characterization of the prior is enormously useful in bringing the final assessment into the regime of the reasonable. Readers are encouraged to experiment with this particular example, varying the weighting of the "0 intake" hypothesis through a range of plausible values, and observe the effect on the posterior distribution.

8.4 CONCLUSION

Bayesian inference combines two sources of information: prior expectations and measurement results. In general, both of these sources of information may be crucial to reaching the conclusion. Bayes's Rule ensures that the conclusion is not overly influenced by either: measurement results will dominate when they are sufficiently numerous and precise to be compelling, but prior expectations will dominate when measurement results are ambiguous.

Subjective information may be used in establishing a suitable prior. Although this raises the possibility of disagreement between different parties, it also allows for the incorporation of vital information that cannot be quantified objectively. Such disagreement is legitimate, in that it reflects actual ambiguity in our knowledge of the problem. Bayesian methods make it possible to proceed to reasonable conclusions even when such ambiguity is present.

8.5 AUTHOR PROFILE

Tom Tadfor Little is a member of the radiological dose assessment team at Los Alamos National Laboratory. His educational background is in physics and astronomy, having graduated from Oxford University in 1985 and received a Ph.D in astronomy from New Mexico State University in 1990. His current areas of interest include Bayesian dose assessment methods and dosimetry of stable tritium compounds.

Chapter 9

USE OF BAYESIAN SOFTWARE TOOLS FOR INTERNAL DOSIMETRY

Guthrie Miller

9.1 INTRODUCTION

The health physicist concerned with internal dosimetry ultimately desires practical tools to apply to the solution of actual problems encountered on the job. The underlying methodology must be considered valid, but this is not the subject of this chapter. The particular application of Bayesian statistics to internal dosimetry considered here is described in detail in Miller et. al. 1999a and 1999b. In this chapter we will use examples of internal dosimetry cases from Los Alamos to demonstrate how Bayesian software tools can be applied to real-life internal dosimetry problems. The Bayesian software package Bayes II for windows systems available from the web site www.lanl.gov/bayesian is used. It is assumed that the reader will have downloaded this package and printed the manual describing the **UF** (Bayesian unfolding) code. In order to reproduce the data plots shown here, the reader needs a copy of the commercial plotting package **ORIGIN**, although other plotting software could also be used to accomplish the same thing.

Five examples are considered. These examples involve the nuclides ^{239}Pu and ^{238}Pu. The method applies equally well to any radionuclide, but a library of biokinetic response functions must be available. The Bayes II package includes biokinetic response functions for ^{239}Pu, ^{238}Pu, and ^{241}Am, using ICRP-30 models and ICRP-60 models. We begin by discussing the UF3.1 code.

9.2 THE BAYESIAN UNFOLDING CODE UF3.1

The programs comprising the Bayesian software package (Bayes II) are shown diagrammatically in Fig. 9.1. The functions of these programs are summarized in Table 9.1. The files shown in Fig. 9.1 are described in Table 9.2.

The urine bioassay data and information about incidents is contained in the input data file for the **UF** code, named **URINE.IN**. The data files for the five examples considered here are called **URINE.111**, **URINE.222**, etc. To calculate for a particular example, copy that urine file into **URINE.IN**. Another **URINE.IN** file, called **URINE.TST,** contains urine bioassay data based on a particular intake scenario (10 nCi intake of class Y 1 μm AMAD ^{239}Pu by inhalation) on a particular date (2/28/1993) using the ICRP-30 biokinetic model. This simulated data is used as a validation tool since the data corresponds to a known intake. This file is shown in Fig 9.2. Using this file as the **URINE.IN** input file, the terminal output when we run the **UF** code (by typing "UF" at the command line) is shown in Fig. 9.3.

The UF output will now be discussed in some detail. The output begins with a listing of the input parameters. If some parameters have not been specified in the input file **UF.IN**, this shows the default values used. A description of input parameters is given in Tables 9.3 and 9.4.

Next in the UF output listing is a message stating that the prior probabilities of various intake types (as listed in the **NAMES.DF** file) are all equal, except that the first is a factor **PFAC** greater. Following is a list of incidents (potential intake dates). These correspond to lines beginning with "I" in the urine data input file **URINE.IN**. The incident date is followed by one or more characters that limit the choice of possible biokinetic models. For example an "i" means only inhalations (biokinetic models whose names start with "i"). An "iw" means only class W inhalations (biokinetic models whose names start with "iw"). Other information that appears on the incident line are the log-normal median $a = \ln(\mu_g)$, the geometric standard deviation $s = \ln(\sigma_g)$ [the mean and standard deviation of $\ln(X)$, where X is the input amount], and the prior probability that intake amount is nonzero (not in a delta function at zero). The statement about intake probability factors means that no lines beginning with an "p" were found in the **URINE.IN** input file that would modify the non-incident probability of an intake as a function of time.

The following lines contain information from the biokinetic model files (those listed in the **NAMES.DF** file). The isotope name, half life, number of body organs for which doses are calculated, and most affected body organ. The "max fractional difference" gives an indication of interpolation errors in using the biokinetic model data tables. If the tables are smaller, the interpolation errors are larger.

Next are reminders that the fit files **FIT0.TXT**, after the first iteration and **FIT.TXT**, after the last iteration, have been written. The "Data with same date combined" statement is a reminder that if two or more bioassay data have the same date, the data are statistically combined. The listed data are the combined resultant data. The (possibly combined) bioassay data are then listed, giving date, measurement result, measurement uncertainty, biological/sample collection variability, "T" or "F", x_{crit}, and

P[$X > x_{crit}$], where X is the intake amount. The value “T” means an incident occurred in the preceding sampling interval, in which case the potential intake date is the incident date. The value “F” means the potential intake date is the midpoint of the preceding sampling interval. The quantity x_{crit} defines “positive" [$X > x_{crit}$] after the first iteration. The quantity P[$X > x_{crit}$] is the prior probability of an intake in the preceding sampling interval using this value of x_{crit}.

The additional information is mostly self explanatory. The units used for the bioassay measurements, the intake amount, and the dose are given in terms of SI units of Bq and Sv. For the example shown, the units are pCi (0.037 Bq) for bioassay measurement, nCi (37 Bq) for intake amount, and rem (0.01 Sv) for dose. These units are specified in the **URINE.IN** file.

The different quantities referred to as “Chisq" in the output listing are 1) “Chisq0", the value of χ^2 assuming no intake, 2) “Chisq", the overall χ^2 calculated using the Bayesian posterior average intake(s) to calculate the “fit”, and 3) the “Chisq" listed with a particular intake, which is the Bayesian posterior expectation value of χ^2 for the block of data used to determine the intake. Recall that χ^2/N, where N is the number of data points, should be about 1 for a statistically satisfactory fit (asymptotically true for N $\rightarrow \infty$, but a useful rough approximation in any case). Smaller than that indicates that the uncertainties used for the bioassay measurements or the biological/sample collection variability are too large, and larger means that the data lies outside the “universe” of possible explanations contained in the family of biokinetic models used, possibly because the data itself are faulty.

The probability of the “most probable" model type referred to in the output listing is the maximum of the Bayesian posterior probability over all the models used to describe this data block. The probability of an intake is the Bayesian posterior probability distribution integrated over all intake amounts larger than $x_{crit.}$ The expectation value of intake amount is then given followed, in parenthesis, with the lower (**EPSL**), middle (50%), and upper (**EPSU**), credible limits of the cumulative Bayesian posterior distribution. The “organ" dose is the committed dose equivalent to the most affected body organ, which is assumed to be the first body organ for which doses are given in the biokinetic model data files.

The biokinetic models used are specified by the file **NAMES.DF** shown in Fig. 9.4. The biokinetic model names are defined in this file and associated with biokinetic model data files. The number of times in the interpolation tables **NTIMES** is arbitrary, although too small a number leads to large interpolation errors. In order to calculate committed doses, the last time value should be at least 18263 days (50 years). It may need to be longer to accommodate very long employment histories over 50 years in length.

A portion of the data file for a particular biokinetic model ("iym_M") is as shown in Fig. 9.5. There are 300 lines of data, giving urine excretion and dose for various times. Only the last few lines are shown. The lines including and after the "format for above:" line are for documentation only. The quantity **NTISSUE** gives the number of tissues for which doses are calculated. The program **UF** uses only the first organ dose. The program **DOSE** uses all **NTISSUE** organ doses to calculate detailed dose information.

Two other data files define the prior probability of an intake for non-incident cases. These are shown in Fig. 9.6.

The **ORIGIN** plot for this example is shown in Fig. 9.7. Fig. 9.8 shows the same data on a log-log scale, where the time variable is now days after intake (incident date) obtained by setting **NI1** equal to 1 (the first incident is used to define time zero) in the input parameter file **UF.IN**.

For this validation case we see that the Bayesian expectation fits the urine data very well, and the calculated dose (CEDE=0.028 Sv) is quite close to the actual value (CEDE=0.0287 Sv).

We now proceed to discussing application of the **UF** code to real plutonium internal dosimetry problems from Los Alamos.

9.3 EXAMPLE 1—NON-INCIDENT-RELATED ^{239}Pu INTAKE DETECTED FROM ROUTINE MONITORING

"Incidents" are defined by work-place indicators, for example air monitor alarms, elevated nose swipe results or high external contamination levels. For plutonium, non-incident related intakes are extremely rare (about 1 in 1000 bioassay samples, see Miller et. al. 1999b), but they do occur. An example is shown in Fig. 9.9, which shows the ^{239}Pu bioassay history from a Los Alamos worker from 1984 to 1998. The curve "expect" is the expectation value of the urine excretion calculated from the Bayesian posterior probability distribution. The Committed Effective Dose Equivalent (CEDE) for the intake shown in Fig. 9.9 is 0.11 ± 0.022 Sv (expected value ± 1SD) . The 5%, 50%, and 95% limits of the posterior distribution are 0.069, 0.11, and 0.14 Sv.

The biokinetic models are drawn from the ICRP-30 set of models as called out in the **NAMES.DF** file, as has been discussed. There is the option to use ICRP-60 models as well. To set up biokinetic model files for a particular nuclide, for example ^{239}Pu using

ICRP60 models, type at the command line **getdfs icrp60 pu239.** The choices are ICRP30, ICRP60 and ^{239}Pu, ^{238}Pu, and ^{241}Am.

Using ICRP-60 rather than ICRP-30, the results are considerably different. The comparison of urine data and expectation value is shown in Fig. 9.10 using ICRP-60 biokinetic models. The Committed Effective Dose is 0.021 ± 0.006 Sv, with 5%, 50%, and 95% credible limits of 0.017, 0.020, and 0.024 Sv.

The expected value is an average over different types of biokinetic response. For ICRP-30, the most likely response type is class Y, large particle size (5 μm AMAD). For ICRP-60, the most probable response function turns out to be type M, large particle size (10 μm AMAD). This accounts for the factor of 5 discrepancy in the dose. For given urine results, ICRP-60 doses for type S are about a factor of 2 less than those from ICRP-30 class Y, while type M (class W) doses are less than those from type S (class Y).

Note that the lower credible limit of the ICRP-30 dose, 0.069 Sv is quite a bit larger than the upper credible limit of the ICRP-60 dose, 0.024 Sv. This is because the calculated uncertainties do not include an uncertainty associated with the family of biokinetic models. This source of uncertainty could be taken into account using the Bayesian method, and it would be reasonable to calculate a posterior probability of ICRP-30 relative to ICRP-60 based on the goodness of fit of models within each family to the urine excretion data. A prior probability could be used to relatively weight ICRP-30 compared to ICRP-60. The calculated uncertainty would then involve an average over different families of models and would be significantly larger.

9.4 EXAMPLE 2—INCIDENT-RELATED ^{239}Pu INTAKE

In this 1993 case there were high-level work place indicators that an intake had occurred, a CAM alarm and nose swipe results of 366 dpm / 814 dpm. The individual had been involved in two previous incidents, but whether or not these incidents had resulted in intakes was not completely clear, although the individual was clearly excreting plutonium from a previous intake or intakes, whether or not these were incident related. The data and ICRP-30 UF interpretation are shown in Fig. 9.11. There are two incident-related intakes in this interpretation of the data. The 1993 intake resulted in a CEDE of 0.14 Sv. If we switch to ICRP-60 biokinetic models, the earlier intakes are not incident related as shown in Fig. 9.12. The expected value curve now has an increased width reflecting the uncertainty of the background caused by the earlier intakes. The 1993 CED becomes 0.051 Sv, still an overexposure.

For the ICRP-60 interpretation, the probabilities of the three intakes are 0.85, 0.596, and 1.00. Normally we would like the intake probabilities to all be nearly 1, and the 0.596 probability is low. The odds that an intake actually occurred are only 0.596/(1.-0.596) = 1.47 to 1. The **UF** code parameter **PROBC** (specified in the **UF.IN** file) defines the lowest probability allowed for an intake. If we change the value of this parameter from 0.5, as it was for Fig. 9.12, to 0.6, the result shown in Fig. 9.13 is obtained. There are now two intakes, both with probability of intake 1.00. The 1993 CED is now 0.062 Sv.

9.5 EXAMPLE 3—MULTIPLE ^{238}Pu INTAKES

In this example there are three intakes, none of them incident related. The urine excretion data and ICRP-30 interpretation are shown in Fig. 9.14. The ICRP-60 interpretation is shown in Fig. 9.15. Remember to type **getdfs icrp30 pu238** to set up the biokinetic model files for ^{238}Pu.

The CEDE's and most probable biokinetic responses are shown in Table 9.5. Note that both ICRP-30 and ICRP-60 models identify the same intake times—a good indication! Although the calculated uncertainties include many components, missing is the uncertainty associated with the nonlinear **UF** code deciding to interpret data with a different intake scenario, in effect saying "no the intake doesn't occur there, it occurs here". An example of this has been already seen in example 2. A full straight forward Bayesian treatment of the internal dosimetry problem rather than an unfolding technique would remedy this situation, as discussed in Miller et. al. 1999a, although it seems presently to be not feasible because of the high dimensional integrations involved.

9.6 EXAMPLE 4—^{239}Pu OVEREXPOSURE DETECTED FROM ROUTINE MONITORING

This case attracted attention recently. The data and ICRP-30 interpretation are shown in Fig. 9.16. The first elevated urine result from December 1998 did not trigger a resample, since the intake probability was only 21%, below the 50% cutoff. Normally high results are followed up by resampling when the intake probability exceeds 50%, and dose results are reported when enough data has been accumulated to have the intake probability be nearly 1. Table 9.6 shows the time progression of our knowledge of this intake after each data point. The expected dose and 5%, 50%, and 95% credible limits are shown. The elevated urine result was obtained by Radiochemical Alpha Spectrometry (RAS), but it was preceded by a higher sensitivity result obtained by Thermal Ionization Mass Spectrometry (TIMS) so the time of intake is bracketed between 19-August-1998 and 15-December-1998. Using ICRP-60, the CED is 0.029 Sv.

We note that up until now the calculated TIMS measurement uncertainties have been somewhat arbitrarily increased over their nominal values using the calculation $\sigma \rightarrow (\sigma^2 + (\sigma_0^2))^{1/2}$, where σ_0 = 15 μBq, pending studies of the distribution of data for urine blanks. We have now implemented a system of continuously tracking the distribution of urine blanks as a quality assurance measure, and such a correction would have σ_0 in the range of 2-4 μBq.

9.7 EXAMPLE 5—^{238}Pu 1999 INTAKE RULED OUT BY PRIOR

In this case an individual who had had a previous intake had two elevated urine results in 1999, seeming to imply another intake in 1999. However the individual had transferred to another job outside the plant in 1994 and as a result of this transfer had a greatly reduced prior probability of intake. The ICRP-30 interpretation shown in Fig. 9.17 has a prior probability factor of 1/10 applied beginning 1-January-1994. Such a modification of the prior probability of an non-incident-related intake is specified in the **UF** code by lines in the **URINE.IN** file beginning with a "p" (as in the **URINE.555** file). With this factor, there is no 1999 intake. Without it (or with the input line commented out), there is a 1999 intake of 0.24 Sv. The higher than normal excretion of plutonium may have been caused by medications the individual was taking, but we have no clear understanding of its cause.

9.8 REFERENCES

Miller, G.; Inkret, W. C.; and Martz, H. F. Internal Dosimetry Intake Estimation Using Bayesian Methods. Radiation Protection Dosimetry, 82:5-17; 1999a.

Miller, G. , Inkret, W. C., Little, T. T. , Martz, H. F. , and Schillaci, M. E. Bayesian Prior Probability Distributions for Internal Dosimetry. Radiation Protection Dosimetry (*submitted*, preprint available for download from www.lanl.gov/bayesian) 1999b.

9.9 AUTHOR PROFILE

Guthrie Miller is a staff member on the radiological dose assessment team at Los Alamos National Laboratory with primary responsibility for transuranic internal dosimetry. Guthrie entered the field of health physics in 1991 coming from a background in high-energy physics and magnetic fusion research. The Los Alamos dose assessment team has published a number of papers on the application of Bayesian statistics to internal dosimetry. Guthrie's web page can be found at http://drambuie.lanl.gov/~guthrie.

SUBJECT INDEX

aleatory, 122
analytical solution, 116, 119
Bayes, 55, 57, 58,59, 61, 63, 66, 69, 721, 73, 76, 77, 79, 80, 81, 83, 84, 86, 315, 316, 319,321, 323
Bayesian inference, 58, 66, 83, 84, 313, 321
Bernoulli process, 89
Binomial distribution, 89
BUGS, 65, 74, 75, 79, 83
Cauchy theorem, 108
CDF, 95, 97, 126, 132, 149
charged particles, 186, 187
Chord-length distributions, 101, 108
Classical estimation, 67
Composite sampling, 296, 309
Compton scattering, 97, 109
Conditional probability, 94
Confidence interval, 1, 176, 177, 179, 191, 226, 227
covariance, 145
Cumulative distribution function, 95, 96
Data Quality Assessment (DQA) process, 273
Data Quality Objectives (DQO) process, 271, 272
distribution, 56, 57, 58, 60, 61, 62, 63, 65, 66, 67, 70, 71, 72, 73, 74, 76, 77, 78, 79, 89, 89, 90, 92, 93, 95, 96, 97, 98, 99, 101, 102, 103, 113, 122, 124, 125, 126, 127, 128, 129, 130, 132, 133, 135, 137, 138, 139, 143, 144, 146, 147, 148, 149, 150, 151, 152, 153, 154, 155, 156, 158, 160, 173, 175, 176, 179, 183, 193, 197, 198, 207, 213, 214, 216, 217, 218, 219, 220, 221, 223, 227, 228, 236, 248, 249, 252, 257, 263, 286, 288, 303, 309, 313, 320, 325, 326, 329
Double Sampling, 294
Empirical Bayes, 76, 77, 84
epistemic, 122
Error, 89, 159, 170, 171, 182, 185, 190, 205, 209
Exponential distribution, 93
Exponential transformations, 112
Forward selection, 192
geometric mean, 139, 140
geometric standard deviation, 126, 140, 324
Gibbs sampling, 76
Hierarchical Bayes, 65, 72
HPD, 71
Hypothesis Tests, 235
Importance sampling, 110, 148, 157
Independent increments, 93
Joint probability, 94
Judgment Sampling, 291
Latin Hypercube, 148
Latin Hypercube sampling, 148
LET, 99, 101, 103, 108, 110
LET unfolding, 108
Likelihood, 63
Linear attenuation coefficient, 92, 97, 112
Linear energy transfer, 99
Linearization, 183
lognormal distribution, 126, 127, 138, 139, 214, 218
Mann-Whitney Test, 239
MAP, 63

Markov chain Monte Carlo, 73, 84
Maximum likelihood, 57
MCMC, 55, 73
Mean, 81, 89, 89, 90, 92, 98-100, 108, 109, 142, 144, 171, 205, 216
Mean Squared Error, 171
microdosimetry, 101, 103, 113
Monte Carlo, 55, 73, 84, 89, 86, 87, 89, 86, 87, 89, 90, 93, 91, 97, 98, 99, 100, 103, 105, 106, 108, 109, 110, 112, 113, 128, 136, 146, 147, 148, 149, 150, 151, 155, 156, 157
Multi-Agency Radiation Survey and Site Investigation Manual (MARSSIM), 271, 310
National Analytical Management Program, 304
Nonlinear regression, 261
normal distribution, 60, 127, 143, 146, 152, 153, 155, 213, 214, 217, 219, 236, 257, 263, 303
Parameter, 121, 122, 149, 154, 203, 205, 209, 279
PDF, 95, 97, 101, 105, 106, 107, 110, 126, 132
Pearson Correlation, 254
Percentiles, 126
Photoelectric absorption, 97, 109
Poisson distribution, 56, 59, 66, 67, 71, 126, 129, 130, 207, 213, 214
Power, 99, 180
Predicted value, 189
Prediction interval, 178, 179, 183, 198
probability, 55, 56, 57, 59, 60, 62, 63, 65, 67, 69, 70, 71, 89, 92, 93, 94, 95, 96, 97, 101, 102, 103, 105, 106, 108, 112, 117, 122, 124, 126, 128, 129, 135, 146, 151, 157, 158, 160, 174, 176, 180, 214, 215, 216, 218, 219, 224, 225, 231, 234, 237, 275, 284, 285, 286, 288, 291, 298, 300, 305, 307, 313, 314, 315, 316, 317, 318, 319, 320, 324, 325, 326, 327, 328, 329
probability density function, 60, 62, 95, 102, 103, 105, 106, 108, 151
probability, Radiation transport, 188, 189
propagation, 141, 156
pseudo-random numbers, 89, 87, 113
Quality Assurance Project Plan, 276, 289, 310
Radiation transport, 186, 187
Random number, 92, 97, 98, 109, 112
Random variable, 89, 90, 95
Ranked-Set Sampling, 295
Regression Sum of Squares, 172
Residual plots, 183
Russian roulette, 112
Sampling, 74, 136, 147, 148, 150, 287, 290, 291, 292, 293, 294, 295, 296, 299, 300, 301, 303, 304, 307, 310
SampTOOL, 304, 305, 307
Scram, 73, 77
simple random, 147, 150, 287, 290, 291, 292, 293, 299, 300, 301, 303, 304
Simple Random Sampling, 291
slope, 216, 228, 261, 262, 263, 264, 265, 266

Spearman Correlation, 254
Standard deviation, 89, 90, 216
Standard error, 177
Stepwise selection, 192
Stratified sampling, 109, 110, 293
Studentized residuals, 198, 199, 200
Subjective probability, 157, 160, 315, 316
subjective probability distribution, 135
Sum of Squared Errors, 169
Systematic Sampling, 293
tolerance interval, 152, 154, 155, 156
Total Sum of Squares, 171
triangular distribution, 127, 133, 155, 158
Type A and B, 122, 123
Type I decision error, 284
Type I error, 117, 285
Type II error, 117, 284, 285, 303
Uncertainty, 117, 122, 123, 136, 141, 155, 156, 157, 159, 160, 311
variability, 56, 67, 80, 115, 121, 122, 123, 124, 128, 129, 133, 135, 149, 150, 154, 155, 157, 158, 171, 172, 178, 183, 189, 191, 192, 206, 220, 244, 293, 299, 302, 324, 325
variance, 60, 63, 78, 79, 109, 110, 123, 125, 129, 130, 136, 137, 139, 142, 143, 144, 145, 146, 148, 150, 151, 152, 153, 154, 168, 173, 177, 181, 182, 184, 185, 199, 200, 209, 210, 248, 250, 298, 299, 300, 303
Visual Sample Plan (VSP), 271, 272, 292, 303, 312
Weighted least squares, 207
Wilcoxon Rank Sum (WRS) test, 287, 301, 303